Lectures on
Cosmology
and
Action at a Distance
Electrodynamics

WORLD SCIENTIFIC SERIES IN ASTRONOMY AND ASTROPHYSICS

Editor: Jayant V. Narlikar
Inter-University Centre for Astronomy and Astrophysics, Pune, India

---

WORLD SCIENTIFIC SERIES IN
ASTRONOMY AND ASTROPHYSICS
Vol. I

Series Editor: *Jayant V Narlikar*

Lectures on
Cosmology
and
Action at a Distance
Electrodynamics

F Hoyle
Bournemouth,
Dorset, UK

Jayant V Narlikar
Inter-University Centre for Astronomy and Astrophysics,
Pune, India

World Scientific
Singapore • New Jersey • London • Hong Kong

Published by

World Scientific Publishing Co Pte Ltd

P O Box 128, Farrer Road, Singapore 912805

USA office: Suite 1B, 1060 Main Street, River Edge, NJ 07661

UK office: 57 Shelton Street, Covent Garden, London WC2H 9HE

Library of Congress Cataloging-in-Publication Data
Hoyle, Fred, Sir.
    Lectures on cosmology and action at a distance electrodynamics /
F. Hoyle, Jayant V. Narlikar.
        p.   cm. -- (World Scientific series in astronomy and
astrophysics ; vol. 1)
    Include bibliographical references.
    ISBN 981022558X. -- ISBN 9810225733 (pbk)
    1. Cosmology.   2. Quantum electrodynamics.   3. Quantum field
theory.   I. Narlikar, Jayant Vishnu, 1938-   . II. Title.
III. Series: World Scientific series in astronomy and astrophysics ;
2.
QB981.H7555   1996
523.1--dc20   96-1986
                                                                    CIP

British Library Cataloguing-in-Publication Data
A catalogue record for this book is available from the British Library.

For photocopying of material in this volume, please pay a copying fee through the Copyright Clearance Center, Inc., 222 Rosewood Drive, Danvers, MA 01923, USA.

Printed in Singapore.

PREFACE

The idea for this book came out of a review article we wrote in 1995 in *Reviews of Modern Physics* to highlight the recent developments in action at a distance electrodynamics. Our article was also meant to be a tribute to the paper by John A. Wheeler and Richard P. Feynman on "Interaction with the Absorber as a Mechanism for Radiation" which had appeared in the same journal fifty years earlier. In this paper Wheeler and Feynman had introduced their now famous absorber theory of radiation in order to present a consistent picture of classical electrodynamics within the framework of action at a distance, that is, without using the notion of a field with its own degrees of freedom.

The absorber theory seemed to hold out hopes of linking the time asymmetry in electrodynamics to the large scale structure of the universe. Unfortunately Wheeler and Feynman did not explore this aspect further and the theory subsequently encountered problems at the quantum level. Later work by others, however, led to further progress on both these fronts with the result that most of the outstanding problems were resolved and by the early 1970s, the action at a distance description became at par with the standard field theoretic one.

Recent developments have taken the theory further and it is now seen that there is indeed a natural cut-off coming from cosmological boundary conditions which eliminates the so-called ultra-violet divergences in quantum electrodynamics. Thus the notorious infinities which require a renormalisation programme are eliminated. With this further input, we believe that the concept of action at a distance is now superior to the conventional field theory.

There is one important caveat, though. The cosmological boundary conditions that produce the right kind of time asymmetry at the classical level and provide the much needed cutoff at the quantum level are not satisfied by the currently popular big bang cosmology. They are satisfied by the steady state and the quasi-steady state cosmology, and in general by cosmologies that have a perfect future absorber and an imperfect past absorber. Moreover, the existence of an event horizon in the future also plays a critical role in making quantum electrodynamics convergent.

The failure of the conventional quantum field theory to deal with the problem of infinities is well known and has remained with us despite numerous attempts to resolve it. The success of the alternative approach described here reminds us that the problem is not purely local but that we need to include cosmological considerations.

We hope that the reader will see the merit of this approach and the important conclusions that it leads to. We urge him or her to weigh its advantages against the sacrifice of preconceived preferences for the big bang cosmology.

This monograph is aimed at students and research workers in theoretical physics and cosmology. We assume knowledge of electrodynamics, quantum mechanics and general relativity, although the description given here is self-contained. We have opted for the "Lecture" format to aim the book at classrooms.

We thank Professor K. K. Phua of World Scientific Publishing Company for inviting us to write this monograph.

Fred Hoyle   Jayant Narlikar
Bournemouth, England   Pune, India

CONTENTS

PART I

CLASSICAL ELECTRODYNAMICS

LECTURE I : HISTORICAL BACKGROUND

A. From Newton to Gauss

The foundations of theoretical physics were laid by Isaac Newton's book *Philosophiae Naturalis Principia Mathematica* published in the mid-1680s. The laws of gravitation and dynamics described therein successfully demonstrated how to explain the various dynamical phenomena ranging from the motions of terrestrial projectiles to the orbits of planets. They also established an important principle: that with suitable initial conditions the subsequent behaviour of a dynamical system can be completely determined provided the forces acting on it are known. Till the advent of quantum mechanics in the early part of this century this deterministic view prevailed.

The next addition to fundamental physics came a century later with the discovery of the electrical force. The law of electrical attraction/repulsion between unlike/like electrical charges as stated by Coulomb was strikingly similar to the inverse square law of gravitation. For a comparison we state Newton's and Coulomb's laws in familiar notation :

$$F_N = -\frac{Gm_1m_2}{r^2}, \tag{1.1}$$

$$F_C = \frac{Ke_1e_2}{r^2}. \tag{1.2}$$

[The constant K can be taken as unity by a suitable choice of units as we shall do hereafter.]

It is possible that Coulomb may have been inspired to think in terms of an inverse square law because of the successes of the law of gravitation. However, the experiments in electrostatics clearly pointed to such a law. Also, in spite of their superficial similarity there was one fundamental difference between the two laws, a difference that led to their subsequent development along different routes. In gravitation there is always attraction whereas in electrostatics the presence of positive and negative charges allows both repulsion and attraction to be present. [Note also that for like charges the rule is of *repulsion* as opposed to *attraction* in gravitation.]

The commonality between the two laws, however, extends beyond the functional (inverse square) form to a deeper level in that they both assume *instantaneous* action at a distance. So far as gravitation was concerned there was no apparant conflict with any observation because of this assumption. In electrodynamics the situation turned out to be different. It became clear as a result of several experiments on rapidly moving charges that the Coulomb law was not sufficient to describe all the observed details. On March 19, 1845 Gauss in a letter to Weber summarized the difficulty in these words (Gauss 1867) :

> ... *I would doubtless have published my researches long since were it not that at the time I gave them up I had failed to find what I regarded as the keystone,* **Nil actum reputans si quid superesset agendum** : *namely, the derivation of the additional forces – to be added to the interaction of electrical charges at rest, when they are both in motion- from an action which is propagated not instantaneously but in time as is the case with light...*

Thus, in a sense Gauss had anticipated the future work of Maxwell but did not get down to the actual description of delayed action at a distance with the speed of light playing the key role. In the post-special relativity era one could express the above requirement that *the action at a distance should be a relativistically invariant concept.* Evidently, with its effect travelling at infinite speed the Newton-Coulomb action at a distance was not consistent with relativity.

B. Maxwell's Field Theory

But, returning to the previous century one can say that action at a distance a-la-Newton-Coloumb-Gauss was sidelined by the success of the field theory of Faraday and Maxwell. While Faraday's approach was experiment-based, Maxwell was able to unify all observed electromagnetic phenomena in a unified formal set of equations. It became easier to describe these phenomena not through direct interaction of electric charges but through a charge ↔ field ↔ charge

type interaction. The 'field', in this picture, has existence of its own.

Thus when charge *1* moves it produces disturbances in the field nearby (see Fig. I.1). These disturbances travel outwards like ripples of a wave and they do so with the speed of light. When they hit charge *2* they make it move. We therefore have the effect of *1* reaching *2* not instantaneously but with the speed of light.

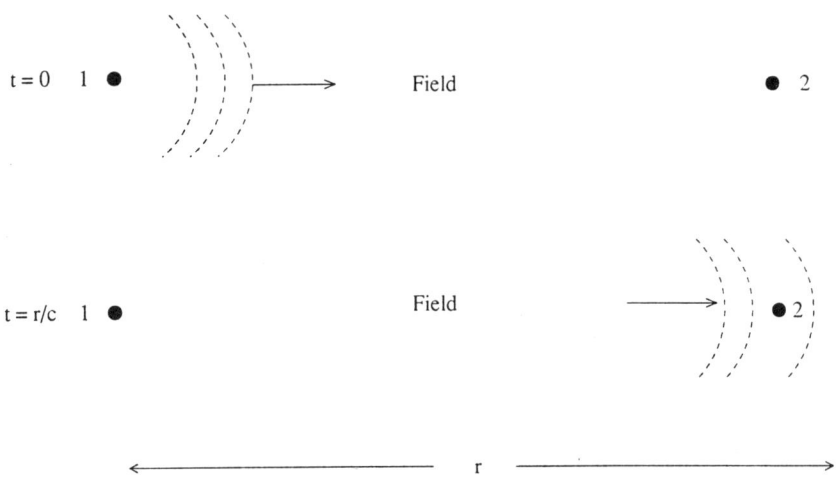

Fig I.1. When electric charge *1* moves it produces ripples in the ambient field. The ripples move outwards with the speed of light. When they reach charge *2*, it experiences the *delayed effect of movement of charge 1.*

In Gaussian units (which we shall use in this book) the Maxwell equations describing charges in an otherwise empty space are given by

$$\nabla.\mathbf{B} = 0 \qquad , \qquad \nabla.\mathbf{E} = 4\pi\rho,$$

$$\nabla \times \mathbf{E} + \frac{1}{c}\frac{\partial \mathbf{B}}{\partial t} = 0 \qquad , \qquad \nabla \times \mathbf{B} - \frac{1}{c}\frac{\partial \mathbf{E}}{\partial t} = \frac{4\pi}{c}\mathbf{j} \qquad (1.3)$$

Here the magnetic induction vector is given by **B** and the electric field by **E**. The (∇.) and (∇×) operators have their usual meanings of divergence and curl. The combination $(\mathbf{E}, \mathbf{B})$ forms the electromagnetic field. The electric charges are represented by their charge density ρ and current density vector $\mathbf{j}$.

If we were dealing with a dielectric and permeable medium with a smoothed out distribution of electric and magnetic dipoles we would have introduced the derived field vectors $\mathbf{D}$ and $\mathbf{H}$ to replace $\mathbf{E}$ and $\mathbf{B}$ in those of Eq. (1.3) which have ρ and $\mathbf{j}$ on the right hand side. We will not explicitly deal with this situation here but refer the reader to standard textbooks in electrodynamics.

The Eq. (1.3) describes first part of the interaction viz. how charges disturb, or act as sources of the electromagnetic field. How does the field disturb a particle of charge e? This second part is described by the Lorentz force

$$\mathbf{F}_L = e\left\{\mathbf{E} + \frac{\mathbf{v} \times \mathbf{B}}{c}\right\}. \tag{1.4}$$

This force acts on the charge as per the laws of motion and makes it move.

These formulae are written in the three dimensional language of Newtonian physics with the implication that time is decoupled from space. Nevertheless, the Maxwell equations held the key to the unification of space and time into a four dimensional spacetime with the Lorentz transformations replacing a combination of Galilean transformations and orthogonal spatial transformations. The Maxwell-Lorentz equations then display a new symmetry and elegance in the four dimensional spacetime.

We will use the four dimensional spacetime notation that became common after special relativity. Thus ($i = 0, 1, 2, 3$) will denote the four spacetime coordinates with $x^0 = ct$ the timelike coordinate and $x^\mu(\mu = 1, 2, 3)$ the three spacelike ones. Here c is the speed of light which on occasions will be set equal to unity to simplify writing. The same will apply to the Planck symbol $\hbar$ which will be needed in our discussions of quantum electrodynamics. In general the latin indices shall take four values 0,1,2,3; while greek indices will take three values 1,2,3. The summation convention shall be assumed. In special relativity the line element is given by

$$ds^2 = \eta_{ik}dx^i dx^k, \tag{1.5}$$

where the metric tensor $\eta_{ik} = diag.(1, -1, -1, -1)$. In general relativity the metric tensor will be denoted by g_{ik}. The line element will continue to have the signature of

Eq. (1.5) even in the latter case although the metric tensor may not be diagonal. A suffix i following the comma denotes differentiation with respect to the coordinate x^i. Likewise, a semicolon denotes covariant differentiation. For the time being we will stick to (1.5).

The field equations are then derivable from an action principle given by

$$J = -\sum_a \int m_a da - \frac{1}{16\pi} \int F_{ik} F^{ik} d^4 x - \sum_a \int e_a A_i da^i. \tag{1.6}$$

In the above the particles $a, b, c, \ldots$ are not interacting directly with one another; they do so through the medium of a field F_{ik} which is defined in terms of a 4-potential A_i by

$$F_{ik} = A_{k,i} - A_{i,k}. \tag{1.7}$$

Following standard convention, the non-zero components of the antisymmetric 4-tensor F_{ik} are given by

$$F_{12} = -F_{21} = B_3, \quad F_{23} = -F_{32} = B_1, \quad F_{31} = -F_{13} = B_2$$
$$F_{14} = -F_{41} = E_1, \quad F_{24} = -F_{42} = E_2, \quad F_{43} = -F_{34} = E_3 \tag{1.8}$$

with $\mathbf{E} \equiv (E_1, E_2, E_3)$ and $\mathbf{B} \equiv (B_1, B_2, B_3)$. The 4-potential A_i likewise can be identified with a combination of the vector potential $\mathbf{A}$ and the scalar potential ϕ. The relation (1.7) is the unified four dimensional statement of the field-potential relationship :

$$\mathbf{B} = \nabla \times \mathbf{A} \quad , \quad \mathbf{E} = -\frac{1}{c}\frac{\partial \mathbf{A}}{\partial t} - \nabla \phi. \tag{1.9}$$

In the four dimensional language the guage condition and the wave equation are also simplified :

$$A^i_{\;;i} = 0, \tag{1.10}$$

$$\Box A_i = 4\pi J_i, \qquad J_i \equiv (\mathbf{j}, \rho c), \tag{1.11}$$

while the Maxwell equations become

$$F_{ik,l} + F_{kl,i} + F_{li,k} \equiv 0, \tag{1.12}$$

$$F^{ik}_{\ ;k} = -4\pi J^i. \tag{1.13}$$

The Lorentz equations of motion of particle a are given by

$$m_a \frac{d^2 a^i}{da^2} = e_a F^i_{\ k} \frac{da^k}{da}. \tag{1.14}$$

Here a^i are the 4-coordinates of particle a and da the element of its proper time.

Thus, Maxwell's field theory not only accounted for the observed phenomena in classical electrodynamics which had led Weber and Gauss to worry about a reformulation of action at a distance but it also proved to be a precursor of special relativity. Subsequently, with the advent of quantum theory, the subject of quantum electrodynamics held the centre stage of theoretical physics in the 1930s and the 1940s and it emerged again with many successes. These include,

1. The explanation of spontaneous transition of an atomic electron

2. The understanding of scattering of electrons, pair creation and annihilation

3. Quantitative confirmation of the Lamb shift and other related effects

With these successes under its belt the Maxwellian electrodynamics not only established a credible picture of the electromagnetic field, but it also made field theory the norm for any basic theory in physics.

However, as we shall see in due course, there are still fundamental problems with the electromagnetic field theory and hence a second look at the earlier alternative – that of action at a distance – becomes opportune. How did this approach fare in the post-Gauss era? Let us take the story further.

C. The Formula For Delayed Action

The problem posed by Gauss was partially solved in the early part of this century by Schwarzschild (1903), Tetrode (1922) and Fokker (1929 a,b, 1932). The concept of action at a distance with action from one particle travelling to the other with the speed of light was eventually expressed by Fokker. We restate below the Fokker formula for delayed action at a distance in a notation that will be useful for describing the subsequent developments.

We begin with the Dirac deltafunction $\delta(x)$ which has the properties:

$$\delta(x) = 0 \quad \text{for } x \neq 0; \qquad \int_{-\infty}^{\infty} \delta(x)dx = 1. \tag{1.15}$$

This satisfies the Dirac identity

$$\eta^{ik}\delta(s_{XA}^2)_{,ik} = \Box_X \delta(s_{XA}^2) = -4\pi\delta_4(X, A) \tag{1.16}$$

where δ_4 is the four dimensional deltafunction for spacetime points $X \equiv (x^i)$ and $A \equiv (a^i)$ and s_{XA}^2 is the square of the interval between them as computed by (1.5).

In Eq. (1.16) $\Box$ is the wave operator and $\delta(s_{XA}^2)$ is its *Green's function*. This identity is valid in the flat spacetime of special relativity and needs to be generalized to the curved spacetime of general relativity which we shall introduce later. For the present we will work within the framework of special relativity.

Having stated our notation we now write the Fokker action formula which describes the interaction between electric charges labelled $a, b, c, \ldots$ etc. as follows:

$$J = -\sum_a \int m_a da - \sum_a \sum_{< b} \int \int e_a e_b \delta(s_{AB}^2)\eta_{ik} da^i db^k. \tag{1.17}$$

In the above expression m_a is the mass of particle a and e_a its electric charge. da is the element of proper time of particle a. The first term is the usual inertial term while the second term is the electrodynamic interaction term. In the latter, the deltafunction ensures that the typical points A and B on the worldlines of a and b interact if and only if they are connectible by a null ray. This is another way of saying that the interaction between A and B propagates with the speed of light. Thus conceptually at least the programme envisaged by Gauss seems to have been achieved. [See Fig. I.2]

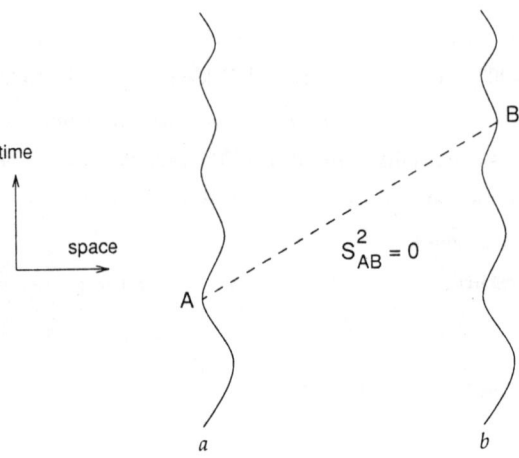

Fig. I.2 The interaction between charges a and b is only between points connectible by a null ray.

But how does it work in practice?

The formula (1.17) looks quite different from the field theory action which was stated in the form (1.6). There the field is an additional, independent entity with its own uncountably infinite degrees of freedom which are called into play in describing phenomena like radiation. What is the corresponding picture in the action at a distance defined by the Fokker formula?

To see the correspondence with the field picture the following definitions of *direct particle potentials* and *direct particle fields* are useful :

$$A_i^{(b)}(X) = e_b \int \delta(s_{XB}^2)\eta_{ik}db^k, \qquad F_{ik}^{(b)} = A_{k,i}^{(b)} - A_{i,k}^{(b)}. \qquad (1.18)$$

Thus we have a field and a potential associated with each particle and these identically satisfy the following relations:

$$A_{;i}^{(b)i} = 0. \qquad (1.19)$$

$$\Box A_i^{(b)}(X) = -F_{i\ ,k}^{(b)\ k} = 4\pi J_i^{(b)}(X) \qquad (1.20)$$

$$J_i^{(b)}(X) = e_b \int \delta_4(X, B)\eta_{ik}db^k. \tag{1.21}$$

Superficially these look similar to the gauge condition, the field equations and the wave equation of the Maxwell field theory. However, these are identities in view of the definitions (1.18). In fact these "fields" do not have degrees of freedom of their own: they are functionals of particle paths. For this reason it is misleading to call them fields. We shall refer to them as *direct particle fields.* In terms of these direct particle fields the variation of the worldline of a typical particle a gives us the analogue of the Maxwell-Lorentz equations of motion :

$$m_a \frac{d^2 a^i}{da^2} = e_a \sum_{b \neq a} F^i{}_k^{(b)} \frac{da^k}{da}. \tag{1.22}$$

Notice that the particle a is acted on by all *other* particles $b \neq a$, i.e., there is no self action. This absence of self action was in fact evident from the Fokker formula which has in the second term the summation excluding self action.

This formulation therefore satisfies the requirement of relativistic invariance and seems to resemble the Maxwellian field theory which is already known as a successful theory of electrodynamics. There are, however, several questions that this formulation has to answer before it can be accepted as a working theory. We list them below.

1. The complete time-symmetry of the formulation tells us that the electromagnetic interaction proceeds not only forward in time but also – in equal strength, it proceeds backwards in time. Fig. 1.1 illustrates this result. Thus there is a manifest violation of causality. How can such a theory explain causal and unidirectional phenomena like radiation?

2. With no degrees of freedom vested in direct particle fields, will the theory be able to account for all electrodynamic observations?

3. How is the theory described in curved spacetime? How does it interact with spacetime geometry? This question assumes significance when we recall that the electromagnetic energy momentum tensor in Einstein's field equations depends entirely on *fields* in Maxwell's theory and that there is no corresponding field term in the present theory.

4. The bulk of the effects of electrodynamics fall within the quantum domain. Can the action at a distance formulation be quantized? Recall again that in the usual formulation it is the field that is quantized and here we have no field.

5. Finally, at a deeper level, we may ask whether this new formulation fares *better* than the standard field theory.

These challenges have been addressed by various workers over a span of several decades. In these lectures we will summarize the progress in the light of the above questions. But let us first go back to the field approach and see why it is still not free from its own fundamental problems. The issues raised above vis-a-vis action at a distance can be better appreciated against the background of the problems faced by the classical field theory of Maxwell. We itemize them in the next lecture under separate heads, although they happen to be interrelated.

Exercises

1. Show that the definitions (1.9) do not automatically guarantee that the gauge condition (1.10) is satisfied. What further transformations of the potential A_i are required to ensure that (1.10) holds?

2. Show that with the definitions (1.18) the guage condition is automatically satisfied if the charge is conserved. What happens if the charge is not conserved?

3. Deduce the Dirac identity (1.16) by taking Fourier transform of the wave equation

$$\Box \phi = -4\pi \delta_4(X, A).$$

4. In what way do the direct particle fields possess fewer degrees of freedom than ordinary fields?

5. Comment on the statement that the direct particle analogues of Maxwell's field equations are identities.

LECTURE II : THE PROBLEMS OF CLASSICAL FIELD THEORY

A. Explanation of Causality

The wave equation (1.11) satisfied by the 4-potential A_i in the Maxwell theory is similar to the relation (1.20) except that in this case it is a genuine equation rather than an identity. Here the right hand side is the current density 4-vector. In terms of our direct particle definition (1.21) it is the sum of all such 4-vectors.

In solving any problem in field theory involving the above equation, it is common practice to choose those solutions of Eq. (1.11) that are consistent with the principle of causality. The most fundamental problem is the one referred to by Gauss (op.cit.), viz. that of the accelerated electric charge. It is well known that the wave equation (1.11) has two independent basic solutions, one having support on the future light cone (the so-called *retarded solution*) and the other having support on the past light cone (the *advanced solution*). Symbolically we will denote these solutions by $A_i^{(ret)}$ and $A_i^{(adv)}$ respectively for the potentials and by $F_{ik}^{(ret)}$ and $F_{ik}^{(adv)}$ for the corresponding fields.

Now in the problem of the accelerated charge, it is customary to select the retarded solution to describe the physical situation. The advanced solution is rejected on the grounds of causality. Thus it is argued that it is physically realistic to have the charge radiating electromagnetic waves which travel outwards from it and reach a distant point at a *later* instant : and the retarded solution describes this situation. The advanced solution describing waves *converging from infinity* onto the source charge and crossing a distant point *before* they reach the source is manifestly unrealistic. Hence the retarded and not the advanced solution is the reasonable one.

While this procedure is entirely consistent with physical reality, at a deeper level it is incomplete; for it does not take us any further towards understanding why the principle of causality should operate. Expressed in a somewhat different form, the phenomenon of radiation by the accelerated electric charge is a unidirectional one in terms of time whereas the basic Maxwell equations are time-symmetric. The question therefore is, *why do we have an electrodynamic arrow of time?* Field theory does not offer any answer. It stops at providing a scenario consistent with causality. The

13

choice of the retarded solution is imposed ad hoc rather than deduced.

B. Radiation Damping

As a result of the choice of the retarded solution and the phenomenon of radiation by the accelerated charge, the charge loses energy and its motion is damped. It is possible to compute the damping force on the charge by using the law of conservation of energy and momentum. In the notation of the preceding section, the equation of motion of a typical charge a is modified from the Maxwell-Lorentz form to the following:

$$m_a \frac{d^2 a^i}{da^2} = e_a F^i{}_k \frac{da^k}{da} + \frac{4e_a}{3} g_{lk} \left(\frac{d^3 a^i}{da^3} \frac{da^l}{da} - \frac{d^3 a^l}{da^3} \frac{da^i}{da} \right) \frac{da^k}{da} \quad , \quad (c = 1). \tag{2.1}$$

The $F^i{}_k$ term here denotes the *external field* acting on the charge. The extra term on the right hand side is the damping force. Notice that it has not been deduced from the basic field theory action whose Lagrangian only gives the Lorentz force. It has been put in from the requirement of energy loss by radiation. For example, if we had chosen a time-symmetric solution, i.e., a solution with half the advanced plus half the retarded fields then there would be no emission of radiation and no damping.

In a highly perceptive discussion of the problem Dirac (1938b) had provided a new modus-operandi for the computation of the force of radiative damping. His prescription was as follows. To the field $F^i{}_k$ used in computing the Lorentz force in Eq.(2.1) add an extra field

$$R^{(a)i}{}_k \equiv \frac{1}{2} \left\{ F^{(a) \ ret \ i}{}_k - F^{(a) \ adv \ i}{}_k \right\}. \tag{2.2}$$

Here $R^{(a)i}{}_k$ is evaluated at the electric charge a. Although both the advanced and retarded fields due to the motion of a diverge on the worldline of a, their difference is finite and as shown by Dirac, its force on the charge is exactly equal to the extra term in Eq. (2.1). Thus the motion of an electric charge a is given by the modified equations :

$$m_a \frac{d^2 a^i}{da^2} = e_a \left\{ F^i{}_k + R^{(a)i}{}_k \right\} \frac{da^k}{da} \tag{2.3}$$

with the second term apparently arising from the charge itself.

The Dirac prescription despite its elegance was somewhat mystifying, however, in that it brought in the advanced solution that had been discarded as unphysical. Dirac sought to relate its presence to another outstanding problem of field theory, namely the problem of infinite self-action. We will consider it next.

C. The paradox of self-action

Dirac (1938b) highlighted the problem with the help of an idealized situation. Imagine an electric charge a at rest and under the action of no forces until it is hit by a hammer. The hit is thus an impulsive force which sets the charge in motion. What happens to the charge thereafter when it finds itself once again under no external forces?

To fix ideas suppose the impulsive force hitting the charge is $P\delta(t)$, in the x-direction, the charge itself being at $x = 0$ at $t = 0$. The charge was at rest in the infinite past, i.e., $\dot{x} = 0$ at $t = -\infty$. Given that a has mass m and charge e, the equation of motion is

$$M\ddot{x} = P\delta(t) + \frac{2e^2}{3c^3}\dddot{x}. \tag{2.4}$$

The second term on the right hand side is the Dirac force of radiation reaction. The equation (2.4) is non-relativistic. Although the relativistic version can be written down, it does not bring any significant new feature into discussion. So we will later refer to it only briefly and continue to work here with Eq. (2.4).

The standard solution of this problem is as follows. We assume (quite naturally, in view of the causality principle) that $\dot{x} = 0$ for $-\infty < t \leq 0-$. Because of the impulsive force, the charge acquires a momentum P at $t = 0+$. Thus it has a starting velocity $\dot{x} = P/m$ at $t = 0+$. Subsequently, $(t > 0)$ the equation of motion drives it along the exponential track :

$$\dot{x} = \frac{P}{m}\exp\left(\frac{3mc^3}{2e^2}t\right). \tag{2.5}$$

This is the so-called 'run-away' solution in which the charge gains unbounded speed in a characteristic time that is a few times

$$\tau = \frac{e^2}{mc^3} \sim 10^{-23} s. \tag{2.6}$$

In the full relativistic case, the electric charge acquires a near-light speed in a time of the order of τ.

How can the charge be so energized? The rationale provided by the field theory is that for a point charge the self-energy of the field is infinite :

$$\lim_{r \to 0} \frac{e^2}{r} = \infty. \tag{2.7}$$

This is the energy reservoir that is tapped as the charge goes onto the run-away solution.

Nevertheless, the solution is peculiar in that, after the force has ceased to act the charge keeps moving with ever-increasing speed. Also, as we shall see next, the correct continuity conditions are not satisfied at $t = 0$. A more natural final state consistent with the dynamics of impulsive forces is one in which the charge has a constant velocity :

$$\dot{x} = \frac{P}{m} \quad \text{for} \quad t > 0. \tag{2.8}$$

This final state, however, is not consistent with the initial state $\dot{x} = 0$ for $t < 0$; for the equation of motion (2.4) is of third rather than second order. The property of the delta function requires $\ddot{x}$, rather than $\dot{x}$ to be discontinuous at $t = 0$. Dirac showed that the correct extension of Eq. (2.8) to $t < 0$ is given by

$$\dot{x} = \frac{P}{m} \exp\left(\frac{3mc^3}{2e^2} t\right), \quad t < 0. \tag{2.9}$$

While formally Eq. (2.9) is the same as Eq. (2.5) its domain of validity is different. The speed $\dot{x}$ as computed by Eq. (2.9) is finite, tending to zero as $t \to -\infty$, as required. The two solutions are illustrated in Fig. II.1.

This second solution is also consistent with the stipulated data but has the advantage over the first solution in having bounded velocity throughout. But it is also peculiar in that according to Eq. (2.9) the electric charge begins to move *even before* being hit by the hammer!

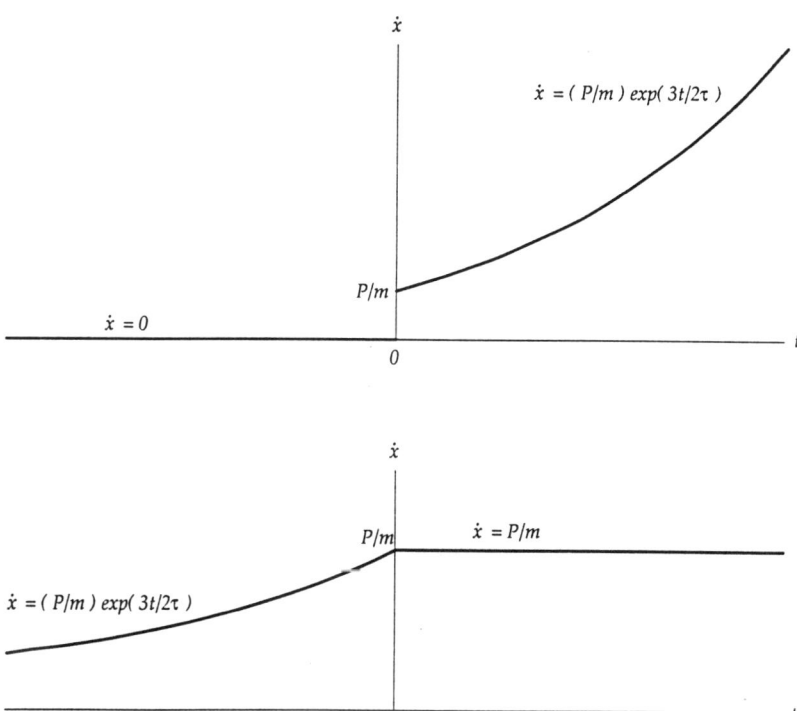

Fig. II.1 The upper figure plots $\dot{x}$ against t for the first solution while the lower figure gives a similar plot for the second solution. In the latter case $\dot{x}$ is continuous at $t = 0$ but $\ddot{x}$ is not. This is the correct requirement at $t = 0$.

This acausal behaviour of the charge can be rationalized by pointing out that the self-action force as computed by Dirac's method does include the advanced field. In practical terms the duration of acausality is of the order τ which is not only very small but also is small by the factor $1/137$ compared to the Compton time scale associated with the charge (assuming that it is an electron or a proton). Thus the problem may well be transferred to quantum theory.

The above discussion (see also Hoyle and Narlikar 1993 for details) offers us a choice between two unphysical solutions. In one we have causality but infinite self-energy while in the other the motions are finite but acausal. Not surprisingly, it was believed that the problem of self-force of the charge would not be solved except by recourse to quantum theory,

This hope has not been fully realized. Quantum field theory does alleviate the self-energy problem but cannot surmount it without introducing the renormalization programme. We shall consider the quantum problem in later lectures. For the present we will confine ourselves to the classical electrodynamics.

These comments therefore underscore the fact that there are conceptual problems with the classical field theory, and thus provide further motivation for looking at the alternative offered by action at a distance.

In 1945 J.A. Wheeler and R.P. Feynman addressed the above issues in an attempt to revive the action at a distance formulation as derived by Schwarzschild, Tetrode and Fokker (see references in Lecture I). The central theme of their argument was that an action at a distance theory was necessarily non-local and that the apparant acausality in its results arose from an inadquate attention being paid to the interaction of a typical charge a with all the other charges in the universe, even if they happen to be located far away. In fact, as they proceeded to demonstrate in a universe that contains enough electric charges to absorb all local electromagnetic influences, the net result is consistent with causality (see : Wheeler and Feynman, 1945).

Exercises

1. Solve the relativistic version of the Dirac problem by following these steps. a) First show that a charge moving under no forces in the x-direction satisfies the

equations of motion

$$\ddot{x} - \tau\dddot{x} - \tau\dot{x}(\ddot{t}^2 - \ddot{x}^2) = 0 \quad , \quad \ddot{t} - \tau\dddot{t}(\ddot{t}^2 - \ddot{x}^2) = 0.$$

where (t, x) are the coordinates in an inertial frame and $c = 1$. The dots are derivatives with respect to the proper time s of the charge.

b) Then show that these equations, with the help of the identity (and its s-derivatives)

$$\dot{t}^2 - \dot{x}^2 = 1,$$

lead to the relation

$$\ddot{t} = \ddot{x}\frac{\dot{x}}{(1 + \dot{x}^2)^{1/2}}.$$

c) Using the first equation of motion in (a) along with the above relation in (b), deduce that for $\ddot{x} \neq 0$,

$$\frac{1}{\tau} - \frac{\dddot{x}}{\ddot{x}} + \frac{\dot{x}\ddot{x}}{1 + \dot{x}^2} = 0.$$

Integrate this to arrive at the parametric solution

$$x = \sinh(e^{s/\tau} + b) \quad , \quad t = \cosh(e^{s/\tau} + b)$$

where b is a constant and the other constant of integration is chosen so that at $s = 0, \tau\ddot{x} = (1 + \dot{x}^2)^{1/2}$.

d) Deduce that as $s \to \infty$ the charge approaches the light-speed barrier.

2. Comment on the energy conservation law for an accelerated electric charge in relativistic or non-relativistic electrodynamics.

LECTURE III : THE WHEELER-FEYNMAN
ABSORBER THEORY OF RADIATION

A. A Simple Illustrative Example

To illustrate how the distant charges influence a local experiment we will repeat briefly the simple derivation given by Wheeler and Feynman in their abovementioned paper.

We assume the universe to be static, Euclidean, with a uniform number density of charges e and with the line element of special relativity as given in Eq.(1.3). Let the charge a be located near the origin O of spherical polar coordinates (r, θ, ϕ) and suppose that its motion there is Fourier analyzed with a typical component of the acceleration given by

$$\mathbf{u}_o e^{-i\omega t}. \tag{3.1}$$

To simplify the picture further, Wheeler and Feynman assumed the local region around the charge a to be empty, in the form of a spherical cavity centred at $r = 0$, and extending as far as $r = R$ and the universe beyond having N charges per unit volume.

In vacuum the full retarded electric field of the charge a at a point P located at a large distance r from it would be given by

$$E_\theta = u_o \frac{e}{r} \sin \theta . \exp\left[i\omega(r - t)\right] \tag{3.2}$$

in the direction of increasing θ where θ is the angle made by the direction OP with that of the acceleration vector $\mathbf{u}_o$. We have taken $c = 1$.

This result, however, needs to be modified to include the refraction effect at the boundary $r = R$ of the cavity and the phase change due to the refractive index $n - ik$ of the medium beyond. The latter is related to the effect the basic field produces on the motion of a typical charge at P. Thus, we modify (2.6) to

$$E_\theta = \frac{2eu_o \sin\theta}{r(1+n-ik)} \exp\left[i\omega\{r-t+(n-ik-1)(r-R)\}\right], \qquad (3.3)$$

and use the field E_θ to compute the acceleration of the charge at P. This is given in the direction of E_θ by

$$\frac{e}{m}p(\omega)E_\theta \qquad (3.4)$$

where $p(\omega)$ is a frequency dependent function, determined in terms of the refractive index by the formula:

$$(n-ik)^2 = 1 - \frac{4\pi Ne^2}{m\omega^2}p(\omega). \qquad (3.5)$$

The crucial step in the Wheeler-Feynman theory was to recognize that in the action at a distance formulation the motion of the particle at P will generate a reaction which will arrive at a backwards in time, i.e., at the instant that the original retarded field left it. This reaction is the half advanced field of the particle at P. Further, to study the electrodynamics in the vicinity of a we must evaluate such responses from all particles lying on the future light cone of a.

The half advanced electric field at a due to the source acceleration at P as given by Eq. (3.2) when resolved in the direction of the acceleration of a then becomes

$$E_\theta \frac{e}{m}p(\omega)\left[\frac{e}{2r}\sin\theta\right]\exp(-i\omega r). \qquad (3.6)$$

The net response of all such particles along the future light cone of a is given by the integral

$$R = \int_{r=R}^{\infty}\int_{\theta=0}^{\pi}\int_{\phi=0}^{2\pi}\frac{e^2}{2mr}p(\omega)\sin\theta.e^{-i\omega r}E_\theta.Nr^2\sin\theta drd\theta d\phi$$

$$= -\frac{2}{3}i\omega u_o e^{-iwt} \qquad (3.7)$$

The responses normal to the acceleration vector cancel out and so we may use Eq. (3.7) to sum over all frequencies and arrive at the result :

$$\mathbf{R} = \frac{2e}{3}\dddot{\mathbf{a}} \tag{3.8}$$

This is the field that the charge a itself would experience because of its action at a distance with the rest of the universe. Multiplying this field by the charge gives us the standard formula for the radiative damping force :

$$\mathbf{R}e = \frac{2e^2}{3}\dddot{\mathbf{a}} \tag{3.9}$$

It can be verified that this is the nonrelativistic version of the Dirac term in Eq.(2.1).

If, instead of calculating the sum of responses at the location of a we had calculated it at an arbitrary point in its neighbouring region, we would have found that the field is Dirac's extra field (2.3),

$$\frac{1}{2}\left\{ F^{(a)ret} - F^{(a)adv} \right\}. \tag{3.10}$$

This calculation is slightly more involved and may be found in the work of Wheeler and Feynman (1945). Using this result, they built up a self-consistent picture of action at a distance in the following way.

In the above calculation the net field emanating from charge a is the full retarded field. How is it made up ? It is made up of two components as given below:

$$F^{(a)ret} = \frac{1}{2}\left\{ F^{(a)ret} + F^{(a)adv} \right\} + \frac{1}{2}\left\{ F^{(a)ret} - F^{(a)adv} \right\}. \tag{3.11}$$

The first term on the right hand side is the basic time-symmetric field of charge a while the second term, as we just saw, represents the response of the universe. The calculation is thus self-consistent since it was the full retarded field that was used in computing the response.

We therefore see that Dirac's mysterious prescription receives a natural derivation in the action at a distance framework. We also see that the radiative reaction is not a self-force but is the combined reaction of the universe to the motion of the charge a. Further, even though our theory is time-symmetric, we seem to have arrived at an explanation of why retarded solutions operate in practice: it is not an *ad hoc* choice required by causality but forced on us by the way the universe responds.

Wheeler and Feynman also showed that it is possible to qualitatively visualize the above derivation in the following way. The typical point P in the universe generates an advanced wave when the retarded wave from the charge a hits it. An advanced wave emanating from P can be visualized as a spherical wavefront from infinity converging on P. Such a wavefront will cross the source charge a before reaching P.

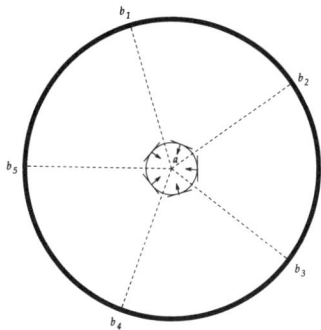

(i) *Wave fronts <u>before</u> crossing source particle $\underline{a}$*

$b_1, b_2, \ldots b_5 \ldots$: *absorber particles*

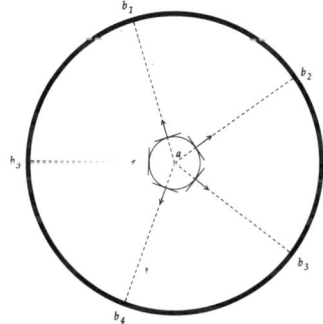

(ii) *Wave fronts <u>after</u> crossing source particle $\underline{a}$*

$b_1, b_2, \ldots b_5 \ldots$: *absorber particles*

Fig. III.1 The waves converging on the absorber particles present almost flat wavefronts as they cross the source particle a. In (i) we see the wavefronts forming a spherical envelope *converging* on a. In (ii) they have all crossed a and are *receding* from it, thus forming a divergent spherical wave. This is how the combination of half retarded minus half advanced waves appears in the radiation reaction formula.

Because of large distance between a and P this wavefront will appear almost plane as it crosses a. Likewise all other particles in the universe will generate similar advancing wavefronts. Since such wavefronts cross a from all possible directions, they simulate a spherical wavefront first converging on a and then diverging away from a. This combination is nothing but the expression (3.10) made up of waves converging at a and diverging from it. Fig. III.1 illustrates this idea.

It might be argued that the elegant result obtained above may be due to our over-simplified choice of parameters describing the universe. The universe is not homogeneous. It may consist of charged particles of various masses (e.g. electrons and protons). The cavity imagined around the charge a may not be spherical. Is the result sensitive to these issues?

Wheeler and Feynman demonstrated that these issues are not important. The crucial issue is that of *complete absorption*. The integral in Eq. (3.7) must then converge to the value it has on the right hand side. This is ensured by the presence of a sufficient number of particles to the future of a that can absorb the disturbance coming out from a and react to it. The condition may be stated thus: *the universe must be a perfect absorber of all electromagnetic fields emanating from within.* The self-consistency argument implied by Eq. (3.11) will not work if the universe is an imperfect absorber. We will now state this requirement mathematically and use it to give a general derivation of the above result.

B. The General Result

Let us consider the universe as static and with Euclidean geometry. The electric charges in it are moving arbitrarily and we will denote by $F^{(a)ret}$ and $F^{(a)adv}$ the retarded and advanced fields of a typical charge a. We remind the reader that the fields referred to here are direct particle fields and hence do not have extra degrees of freeedom of their own. Thus the retarded/advanced field implied here is well defined with respect to the light cones future/past of the corresponding particle.

We now state the property of perfect absorption as implied by Wheeler and Feynman (1945) as follows: When any arbitrary electric charge a is accelerated, all electromagnetic fields arising from its motion - directly or through its interaction with other charges should tend to zero sufficiently rapidly at great distances from a. If we confine our attention to only such fields, then the above condition means

$$\sum_b \frac{1}{2}\left[F^{(b)ret} + F^{(b)adv}\right] \sim o\left(\frac{1}{r}\right) \; as \; r \to \infty. \tag{3.12}$$

In vacuum, a radiative field falls asymptotically as r^{-1} and the more rapid fall implied by Eq.(3.12) indicates perfect absorption. The proof given by Wheeler and Feynman that in a perfectly absorbing universe only retarded interactions survive is as follows.

Since in (3.12) we have a combination of incoming and outgoing waves, for the relation to hold at all times we need the two types of waves to vanish asymptotically *separately*. Thus Eq. (3.12) implies two relations :

$$\sum_b \frac{1}{2}F^{(b)ret} \sim o\left(\frac{1}{r}\right), \quad \sum_b \frac{1}{2}F^{(b)adv} \sim o\left(\frac{1}{r}\right) \quad as \; r \to \infty. \tag{3.13}$$

and hence also the following :

$$\sum_b \frac{1}{2}\left[F^{(b)ret} - F^{(b)adv}\right] \sim o\left(\frac{1}{r}\right) \quad as \; r \to \infty. \tag{3.14}$$

However, unlike Eq.(3.12) the above combination represents a sourceless field and hence a solution of the homogeneous wave equation. As such, its faster than r^{-1} behaviour at infinity implies that it must vanish identically everywhere. Hence

$$\sum_b \frac{1}{2}\left[F^{(b)ret} - F^{(b)adv}\right] \equiv 0. \tag{3.15}$$

The field acting on charge a therefore becomes

$$\sum_{b \neq a} \frac{1}{2}\left[F^{(b)ret} + F^{(b)adv}\right] = \sum_{b \neq a} F^{(b)ret} + \frac{1}{2}\left[F^{(a)ret} - F^{(a)adv}\right]. \tag{3.16}$$

The first term on the right hand side represents the retarded field of all other charges $b \neq a$ acting together on a while the second term is the Dirac radiative reaction.

We therefore arrive at the general version of the result derived in the simple example considered earlier, thus highlighting the role of perfect absorption by the universe. For this reason, Wheeler and Feynman called this theory the *absorber theory of radiation*.

C. Enter Cosmology

The apparant resolution of the causality problem in action at a distance was, how-ever, not quite complete in its logical framework as Wheeler and Feynman themselves pointed out. We can see the problem in the following way. In the above general argu-ment interchange the words advanced and retarded to find that the chain of reasoning still goes through with (3.16) replaced by

$$\sum_{b\neq a} \frac{1}{2}\Big[F^{(b)ret} + F^{(b)adv}\Big] = \sum_{b\neq a} F^{(b)adv} + \frac{1}{2}\Big[F^{(a)adv} - F^{(a)ret}\Big].\qquad(3.17)$$

This means that the charge a is acted on by the advanced fields of all other charges and a radiative reaction that is the exact opposite of that given by Dirac!

There is nothing to prevent us from using Eq.(3.17) instead of Eq. (3.16), but in practice it would be very awkward. For example, in the simple example discussed earlier, we had assumed that the absorber particle was at rest before being hit by the retarded wave from the source. This is reflected in the first term on the right hand side of Eq. (3.16) which is uncorrelated with the motion of a. In Eq. (3.17) on the other hand the first term is highly correlated with the motion of a and hence if we took those correlations into account we would recover Eq.(3.16).

On the other hand we could use Eq. (3.17) to describe a new situation in which the universe admits only the advanced solutions. The simple example discussed earlier in this lecture would then have a counterpart in which the acceleration of a generates advanced, i.e. incoming waves which hit the absorber particles *before* reaching the source charge. A typical absorber particle must move in such a way that after the incoming wave has hit it, it comes to rest.

Wheeler and Feynman argued that while *a-priori* there is nothing to prevent us from imagining a universe with initial conditions set up in the above fashion, in the thermodynamic context such artificial initial conditions would appear highly unlikely. In fact, this distinction between Eqs (3.16) and (3.17) was, according to them, dic-tated by thermodynamic time-asymmetry. The time-asymmetry in electromagnetic radiation arises from asymmetrical initial conditions that favour Eq. (3.16) over Eq. (3.17), i.e., from the time-asymmetry in thermodynamics.

It turns out, however, that this recourse to thermodynamics is unnecessary. The crucial consideration that breaks the time-symmetry of the action at a distance theory comes from cosmology. This was first pointed out by Hogarth (1962) who argued that, if due note is taken of the cosmological fact that the universe is expanding,then the symmetry between the two situations leading to Eqs (3.16) and (3.17) is broken. For, if we examine the proof of the general result of Wheeler and Feynman given above, we notice that to prove the consistency of retarded solutions, we require perfect absorption in the future, and likewise we need perfect absorption in the past for the consistency of the advanced solutions.

The early observations of Hubble (1929) based on the redshifts of the nearby galaxies and clusters have since been extended to galaxies considerably farther away and the picture of the expanding universe has come to be generally accepted. Most cosmological models today are based on this concept. Thus the assumption of a static universe by Wheeler and Feynman was unrealistic. Hogarth's argument can be reworded in the following way to underscore the crucial role of cosmology in action at a distance electrodynamics.

Suppose that we have a universe that has future and past absorbers operating at different efficiencies which we shall denote by factors f and p. Thus $f = 1$ denotes a perfect future absorber while $p = 1$ a perfect past absorber. Let such a universe lead to a net self-consistent solution of the form

$$F_{\text{total}} = AF^{(ret)} + BF^{(adv)} \tag{3.18}$$

where A and B are constants. Now, we have seen that a full retarded solution gives the Dirac radiative reaction in a perfectly absorbing universe. With an absorber of efficiency f the field $AF^{(ret)}$ will therefore generate a radiative reaction Af times the Dirac value. Similarly, the field $BF^{(adv)}$ will generate a reaction $-Bp$ times the Dirac value. For self consistency therefore, the net radiative reaction $(Af - Bp)$ times the Dirac value added to the basic elementary field of the charge a should give us the net field assumed in Eq. (3.18) :

$$F_{\text{total}} = \frac{1}{2}\left[F^{(ret)} + F^{(adv)}\right] + \frac{1}{2}(Af - Bp)\left[F^{(ret)} - F^{(adv)}\right]. \tag{3.19}$$

Equating the coefficients of the advanced and retarded fields in Eqs. (3.18) and (3.19) separately, we determine the coefficients A and B as

$$A = \frac{1-p}{2-f-p} \quad , \quad B = \frac{1-f}{2-f-p}. \tag{3.20}$$

Notice that if $f = 1$ we get the full retarded field as the self-consistent answer so long as $p \neq 1$. Similarly, for $p = 1$ and $f \neq 1$ we get the full advanced field as the self-consistent answer. Only for the case $p = f = 1$ do we run into an ambiguous situation. This last was precisely the case encountered by Wheeler and Feynman. Hogarth on the other hand showed that most familiar cosmological models lead to unambiguous results.

Before we can examine the cosmological implications in detail, however, we first prepare the groundwork for describing action at a distance in curved spacetime, since cosmology uses that framework. Next we discuss a few important cosmological models.

Exercises

1. Show that the absorber theory of Wheeler-Feynman when applied to a static Euclidean universe, produces resultant propagation of radiation along one set of light cones, without really distinguishing between the past light cone and the future light cone.

2. Show by explicit calculation that the Wheeler-Feynman derivation of §**A** works even if the number density is not homogeneous.

3. Suppose the universe has a near-perfect future absorber that allows only a small fraction η of the emitted radiation to escape. If its past absorber is very imperfect ($p \neq 1$ by a wide margin) then the local electrodynamics will have a fraction $\sim \eta/(1-p)$ of advanced radiation.

4. Discuss why Wheeler and Feynman had to invoke thermodynamics to explain radiation asymmetry in electrodynamics.

LECTURE IV : ACTION AT A DISTANCE IN CURVED SPACETIME

A. The Basic Formalism

As discussed above we will first develop the general framework for describing the action at a distance electrodynamics in Riemannian spacetime and then apply it to some of the standard models of the universe. Such a framework was first given by Hoyle and Narlikar (1964a) in their attempts to follow up Hogarth's lead in a more comprehensive manner.

Thus instead of the Minkowski line element of Eq.(1.5) we now have

$$ds^2 = g_{ik}dx^i dx^k \tag{4.1}$$

and the question arises, in what way can we generalize the $\delta(s_{AB}^2)$ type of interaction to the above spacetime. Although the square of the interval s_{AB}^2 between two world points A, B along the geodesic joining them (assuming it to exist and to be unique) is definable in a Riemannian spacetime, any operations of calculus on it are extremely intricate and do not lead us to Maxwell-like equations. The correct procedure lies in the generalization of the Dirac identity (1.16) to curved spacetime. In what follows we will use the standard notation and jargon of general relativity found in any standard text on the subject [see Narlikar, 1979 for example].

Synge (1960) had developed the necessary basic framework which was subsequently used by Dewitt and Brehme (1961) for defining the Green's functions of the wave equation in a Riemannian spacetime. In formulating action at a distance these Green's functions play the basic role of the above deltafunction. We will use the notation of Hoyle and Narlikar (op.cit) in what follows.

Accordingly we will rewrite the Fokker action (1.17) in curved spacetime in the following form:

$$S = -\sum_a \int m_a da - \sum_{a < b} \sum 4\pi e_a e_b \int \int G_{i_A i_B} da^{i_A} db^{i_B}. \tag{4.2}$$

Here the first term is a straightforward generalization but the second one needs some explanation. The Green's function $G_{i_A i_B}$ takes the place of the flat spacetime term $\delta(s_{AB}^2)\eta_{ik}$ and has the following properties:

$$G_{i_A i_B} = G_{i_B i_A} \tag{4.3}$$

$$\Box_X G_{i_X i_B} + R_{i_X}{}^{l_X} G_{l_X i_B} = [-\bar{g}(X,B)]^{-1/2} \bar{g}_{i_X i_B} \delta_4(X,B). \tag{4.4}$$

Notice first that we have attached suffixes to the tensor indices to indicate the space-time point at which they operate. This is necessary since the property of tensor covariance is a local one in curved spacetime. Instead of functions of one spacetime point common in field theory, here we are forced to use quantities that are invariant or covariant at two points where they are defined.

The relation (4.3) indicates the property of symmetry between the two points A, B at which $G_{i_A i_B}$ acts as a vector. It is this property that ensures time-symmetry of the action defined above. The relation (4.4) is the generalized Dirac identity satisfied by the Green's function. The two-point vector on the right hand side is the so-called *parallel propagator* introduced by Synge (1960) to describe the parallel propagation of a vector along the geodesic joining two points.

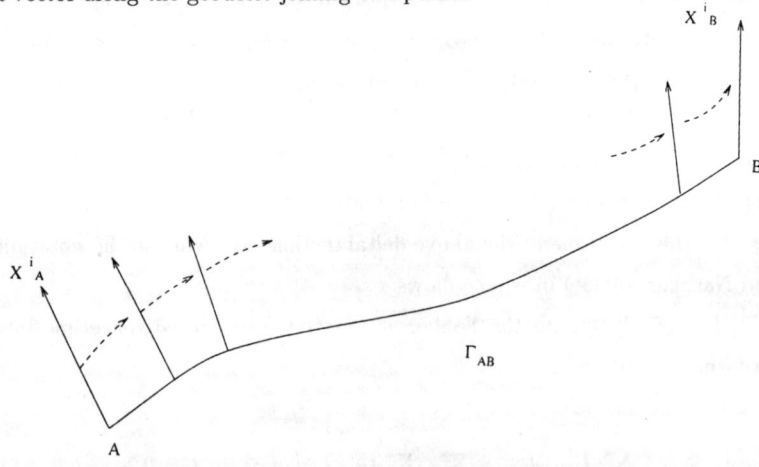

Fig. IV.1 The vector X^{i_A} at A is parallelly propagated to B along the geodesic Γ_{AB} connecting A to B. At B the four components X^{i_B} of this vector will be linearly dependent on the original from components X^{i_A}.

The parallel propagators are defined in the following way. Let A and B be two points in spacetime joined by a (supposedly unique) geodesic Γ_{AB} (see Fig. IV.1). Let an arbitrary vector X^{i_A} at A be parallely transposed to B where its value is denoted by X^{i_B}. Then we write

$$X^{i_B} = \bar{g}^{i_B}{}_{i_A} X^{i_A} \tag{4.5}$$

where $\bar{g}^{i_B}{}_{i_A}$ is a two-point vector, i.e., a vector with index i_B at B and index i_A at A, and is called the parallel propagator. The quantity $\bar{g}$ in Eq. (4.4) is defined by

$$\bar{g}(A, B) = \det \|\bar{g}_{i_A i_B}\|. \tag{4.6}$$

It is not difficult to show that

$$\lim_{B \to A} \bar{g}_{i_A k_B} = g_{i_A k_A} \equiv (g_{ik})_A. \tag{4.7}$$

For detailed properties of $G_{i_A i_B}$ see Dewitt and Brehme (1961) and Hoyle and Narlikar (1964). For example, the Green's function has the following structure

$$G_{i_A i_B} = p_{i_A i_B} \delta(s_{AB}^2) + q_{i_A i_B} \theta(s_{AB}^2) \tag{4.8}$$

where $p_{i_A i_B}$ and $q_{i_A i_B}$ are two-point functions and θ is the Heaviside function. The latter part of the Green's function vanishes in flat spacetime. We therefore see a connection with the $\delta(s_{AB}^2)$ term of the Fokker action. Here we briefly run through the formalism just to illustrate how the action at a distance is describable analogously to the flat spacetime version. Thus the direct particle potential and field are defined by

$$A_{i_X}^{(b)} = 4\pi e_b \int G_{i_X i_B} db^{i_B}, \qquad F_{i_X k_X}^{(b)} = A_{k_X; i_X}^{(b)} - A_{i_X; k_X}^{(b)}, \tag{4.9}$$

and, in view of Eq. (4.4) the following relations also hold:

$$A^{(b)i}{}_{;i} \equiv 0, \qquad F_i^{(b)k}{}_{;k} = -4\pi J_i^{(b)} \tag{4.10}$$

where the current vector is defined as a straightforward curved space analogue of Eq.(1.21). (Where there is no ambiguity the suffix on a tensor index is dropped.) The Lorentz force equation is likewise a generalization of Eq.(1.22) which need not be explicitly stated.

B. The Energy-Momentum Tensor

The action so formulated answers the question 3 raised in Lecture I, even in relation to the effect of direct particle field on the spacetime geometry. For, a variation of the metric tensor alters the spacetime in which the Green's function $G_{i_A i_B}$ is defined. As a result the Green's function also changes and hence the action. It can be shown that this variation leads to an energy momentum tensor defined in terms of direct particle fields that resembles the energy momentum tensor of the Maxwell field theory. The calculation is somewhat tedius and will take us away from the main point at issue. The details may be found in our earlier work (Hoyle and Narlikar 1964a).

Earlier Wheeler and Feynman (1945) had speculated whether the action at a distance theory would produce such a gravitational effect. From purely flat spacetime arguments they had arrived at two possible forms of the energy tensor:

$$T^{ik}_{\text{Frenkel}} = \frac{1}{4\pi} \left\{ \frac{1}{2} g^{ik} \sum_{a \, < \, b} \sum F^{(a)}_{lm} F^{(b)lm} - \sum_{a \, < \, b} \sum \left[F^{(a)il} F^{(b)k}_{l} + F^{(b)il} F^{(a)k}_{l} \right] \right\}$$

$$(4.11)$$

and,

$$T^{ik}_{\text{canonical}} = \frac{1}{4\pi} \left\{ \frac{1}{4} g^{ik} \sum_{a \, < \, b} \sum \left[F^{(a)adv}_{lm} F^{(b)ret \; lm} + F^{(b)adv}_{lm} F^{(a)ret \; lm} \right] \right.$$

$$\left. - \sum_{a \, < \, b} \sum \left[F^{(a)adv \; il} F^{(b)ret \; k}_{l} + F^{(b)adv \; il} F^{(a)ret \; k}_{l} \right] \right\} \qquad (4.12)$$

The first one they called the *Frenkel tensor* and the second the *canonical tensor*. They had concluded :

...From the standpoint of pure electrodynamics it is not possible to choose between the two tensors. The difference is of course significant for the general theory of relativity,

where energy has associated with it a gravitational mass. So far we have not attempted to discriminate between the two possibilities by way of this higher standard...

It was subsequently shown by Narlikar (1974) that it is the canonical tensor that arises from the above variational procedure.

C. Conformal Invariance

We end this lecture by mentioning an important property of electrodynamics, viz., that it is *conformally invariant.* A conformal transformation is specified by a relation between two metrics g_{ik} and $\bar{g}_{ik}$ defined on the same spacetime manifold :

$$g_{ik} = \Omega^2 \hat{g}_{ik}, \tag{4.13}$$

where Ω is a general function of spacetime coordinates, satisfying the relation $0 < \Omega < \infty$. Under this transformation the Maxwell equations (4.10) remain unchanged. For,

$$
\begin{aligned}
J_i^{(b)}(X) &= e_b \int \delta_4(X, B) g_{ik} (-\bar{g})^{-1/2} db^k \\
&\quad - e_b \int \delta_4(X, B) \Omega^{-2} \hat{q}_{ik} (-\hat{g})^{-1/2} db^k \\
&= \Omega^{-2} \hat{J}_i^{(b)}(X),
\end{aligned}
\tag{4.14}
$$

while

$$\hat{F}_{ik} = F_{ik}, \tag{4.15}$$

$$
\begin{aligned}
\hat{F}_i{}^k{}_{;k} &= \hat{g}_{il} \hat{F}^{lk}{}_{;k} = \hat{g}_{il} (-\hat{g})^{-1/2} \frac{\partial}{\partial x^k} [(-\hat{g})^{1/2} \hat{F}^{lk}] \\
&= \Omega^{-2} g_{il} (-g)^{-1/2} \frac{\partial}{\partial x^k} [(-g)^{1/2} F^{lk}] \\
&= \Omega^{-2} F_i{}^k{}_{;k}.
\end{aligned}
\tag{4.16}
$$

Likewise, the other Maxwell equations

$$F_{ik;l} + F_{kl;i} + F_{li;k} = 0 \tag{4.17}$$

can also be shown to be conformally invariant

The relation (4.15) implies that under the conformal transformation, the potentials undergo a guage transformation. That is,

$$\hat{A}_i = A_i + \phi_{;i} \tag{4.18}$$

where ϕ is a scalar function of spacetime. The Green's functions will change accordingly :

$$\hat{G}_{i_A i_B} = G_{i_A i_B} + H_{;i_A i_B}, \tag{4.19}$$

where H is a biscalar.

Although this property of conformal invariance has been noted for Maxwellian electrodynamics, it has special significance in action-at-a-distance theories. For, in such theories the light cones, being the supports for the non-local interaction enjoy a special status. And these light cones are globally invariant under the conformal transformation (4.13). Conversely, if two metrics produce the same family of light cones in a given spacetime manifold, then they are conformally invariant.

Apart from this aspect we will shortly discover that the property of conformal invariance considerably simplifies the discussion of the Wheeler-Feynman theory in cosmological spacetimes.

We now leave these formal aspects of action at a distance in curved spacetime since we shall need them only marginally. Our aim has been to demonstrate that with the above framework it is legitimate to talk of an absorber theory in the curved spacetime of an expanding universe.

Exercises

1. Show that under a conformal transformation, the wave equation

$$\Box\phi + \frac{1}{6}R\phi = 0$$

is invariant. How does ϕ transform?

2. Show that null geodesics do not change under a conformal transformation.

3. Suppose a symmetric Green's function $G(A, B)$ satisfies the scalar wave equation

$$\Box_X G(X, B) = [-\bar{g}(X)]^{-1/2}\delta_4(X, B).$$

Show that if the metric tensor is varied from g_{ik} to $g_{ik} + \delta g_{ik}$ over a region v, then $G(A, B)$ changes by

$$\delta G(A, B) = -\int_v \delta[(-g)^{1/2}g^{ik}]G^{\text{ret}}(A, X)_{,i}G^{\text{adv}}(X, B)_{,k}d^4x.$$

Note that A and B need not lie in v.

4. In Q.3 what condition must A and B satisfy in order that $\delta G(A, B) = 0$?

5. Set up the equations which determine the parallel propagators along null geodesics connecting two points in a conformally flat spacetime.

LECTURE V : COSMOLOGICAL MODELS

A. The Robertson–Walker Spacetimes

For completeness it is now necessary to describe the cosmological models in which in the following subsection we discuss the absorber theory of radiation. Although there are several cosmological approaches we will restrict our attention to those normally described within the metric theories of gravitation. Again, we will limit our discussion to those aspects that we shall need for this article. The reader may refer to standard texts in cosmology for details. Here we will use the notation of Narlikar (1993).

To begin with, we will consider only those models that require the universe to be homogeneous and isotropic on a large enough scale. Such models are described by the Robertson-Walker line element which in standard notation is

$$ds^2 = dt^2 - Q^2(t)\left[\frac{dr^2}{1 - kr^2} + r^2(d\theta^2 + sin^2\theta d\phi^2)\right]. \tag{5.1}$$

Here (r, θ, ϕ) are the comoving coordinates of a typical galaxy which ideally is presumed to be at rest in the above expanding cosmological frame. The time coordinate t is called the *cosmic time* and such a global time coordinate can be defined because of the large scale symmetry (homogeneity and isotropy) assumed for the spacelike sections $t =$ constant. These symmetry arguments allow the spacelike sections to have three types of geometry, all with *constant curvature.* The parameter k denotes the type of curvature : thus for $k = 0$ we have flat Euclidean sections, for $k = +1$ we have closed sections while for $k = -1$ we have open hyperbolic sections. All these sections have an overall scale that varies with epoch.

The scale factor itself is given by $Q(t)$. As we shall shortly see, observations extending over several billion years along the past light cone indicate that the universe has been expanding, i.e., the function $Q(t)$ has been increasing with time over that timespan. Different cosmological models have, however, specified different functional forms for the scale factor. Thus in some cases the universe expands from a singular (pointlike) beginning, the so-called *big bang*, and either expands for ever or contracts back to a pointlike singularity (the *big crunch*). There are also models which are nonsingular.

36

For our purpose we do not need the specific details of gravity theories that lead to these models. As we shall see in the next lecture, we need the geometrical quantities $Q(t)$ and k of the Robertson-Walker model and the way the density of matter $\rho(t)$ falls off at asymptotic past and future. The following Table I gives these details for some well-known models.

Table I

Some Important Cosmological Models

Model	k	$Q(t)$	$\rho(t)$	Reference
Einstein-de Sitter	0	$t^{2/3}$	$\propto Q^{-3}$	Einstein and deSitter (1932)
Closed Friedmann	+1	$A(1 - \cos\psi)$, $A = \text{constant}$ $t = B(\psi - \sin\psi)$	$\propto Q^{-3}$	Friedmann (1922, 1924)
Open Friedmann	−1	$B(\cosh\psi - 1)$, $B = \text{constant}$ $t = B(\sinh\psi - \psi)$	$\propto Q^{-3}$	op. cit.
Steady-State	0	$exp\ Ht$	constant	Bondi and Gold (1948), Hoyle (1948)
Quasi-Steady-State	0	$exp\ At.(1 + \alpha\cos\beta t)$ A, α, β constant	$(1 + \alpha\cos\beta t)^{-3}$	Hoyle, Burbidge and Narlikar (1993)
Brans-Dicke	0	t^A, $A = (2\omega + 2)/3\omega + 4)$	$\propto Q^{-3}$	Brans and Dicke (1961)
Dirac	0	$t^{1/3}$	$\propto Q^{-3}$	Dirac (1938a)

B. The Past and Future Absorbers

Despite differences in their geometrical details these models share certain common features which we now highlight. First, the redshift. If the source of light being observed now, at epoch t_o shows the wavelength of a certain spectral line to be λ, then the wavelength of that line at the epoch t_e of emission was λ_e where

$$1 + z \equiv \frac{\lambda}{\lambda_e} = \frac{Q(t_o)}{Q(t_e)}. \tag{5.2}$$

The parameter z is called the redshift. For $z > 0$, the line has shifted towards the red end of the spectrum. This is invariably the case and so we conclude that observations of all discrete extragalactic sources show that the length scale of the universe has increased since the time light left a typical source.

The models described here predict definite relations between redshift and distance D of an object located beyond our galaxy. For small distances the relationship is linear, being given by

$$z = \frac{H_0}{c}D \tag{5.3}$$

where H_0 is a constant, known as Hubble's constant. The observations of E. Hubble in 1929 first revealed such a linear law for nearby galaxies (see Fig. V.1). It can be shown that $H_0 = \dot{Q}/Q$ evaluated at the present epoch.

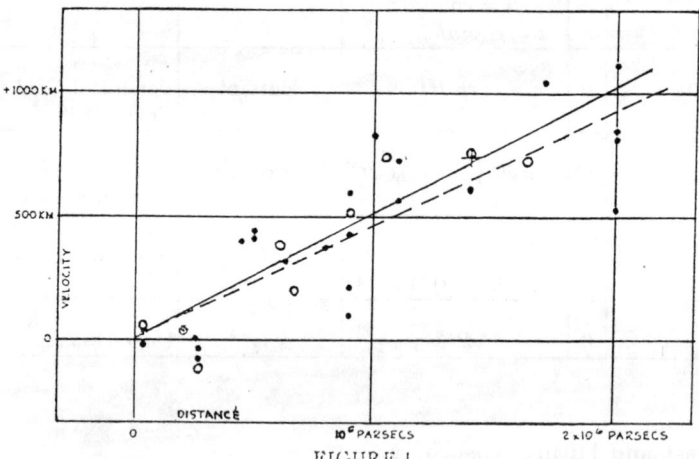

FIGURE 1

Velocity-Distance Relation among Extra-Galactic Nebulae.

Fig. V.1 This graph is taken from Hubble's original paper of 1929. After correcting for systematic errors, the Hubble constant today appears to be ~ 10–15 percent of Hubble's original value.

Thus in an expanding universe, light travelling towards the future (as in a retarded solution) is *redshifted*, while that travelling towards the past (as in an advanced solution) is *blueshifted*. Hence, a future absorber in such a universe will encounter radiation at low frequencies whereas the past absorber will have to deal with high frequency radiation.

A second feature of expanding models different from flat spacetime is the epoch dependence of density. As seen from Table I the density in general behaves as a function of the epoch and thus the past absorber is physically different from the future absorber. Hence, unless one carries out explicit computations one cannot decide how one absorber will respond based on the knowledge of how the other does.

Thus it is clear that when discussing the interaction with the future absorber we are dealing with low energy waves while in the case of the past absorber the interacting waves are of high energy. Likewise, except in the case of the steady state and the quasi-steady state models, and the closed Friedmann model,the future absorber has low density and the past absorber high density of absorbing matter. These issues will be relevant to our discussion of the absorber theory in these models.

C. Conformal Flatness of Robertson–Walker Spacetimes

A fortunate circumstance simplifies the discussion of the absorber theory in the above cosmological models. This arises because (a) these cosmological models are conformally flat and (b) the electrodynamic equations are conformally invariant. The latter aspect we have already noted in the last lecture. To demonstrate (a) we proceed as follows.

If we can find a conformal function and a set of coordinates such that $\bar{g}_{ik} = \eta_{ik}$ then the spacetime described by the metric is said to be *conformally flat*. It was shown by Infeld and Schild (1945) that the Robertson-Walker model is conformally flat. The following series of transformations are needed to explicitly demonstrate this result :

$$k = 0 \qquad \tau = \int_0^t \frac{du}{Q(u)} \quad , \quad \rho = r.$$

$$k = |1 \qquad T = \int_0^t \frac{du}{Q(u)} \quad , \quad r = \sin R,$$

$$\xi = \frac{1}{2}(T + R) \quad , \quad \eta = \frac{1}{2}(T - R),$$

$$\tau = \frac{1}{2}(tan\ \xi + tan\ \eta), \quad \rho = \frac{1}{2}(tan\ \xi - tan\ \eta).$$

$k = -1$ Same as above with hyperbolic functions replacing trignometric ones.

$$ds^2_{R-W} = \Omega^2[d\tau^2 - d\rho^2 - \rho^2(d\theta^2 + sin^2\theta d\phi^2)]. \tag{5.4}$$

We will give below the explicit examples of the Einstein deSitter model and the steady state model.

(i) Einstein deSitter Model :

$$Q(t) = \left(\frac{t}{t_0}\right)^{2/3}, \quad t_o = \text{constant}, \quad \tau = 3t_o^{2/3}t^{1/3}, \quad \tau_o = 3t_o$$

$$\Omega(\tau) = \left(\frac{\tau}{\tau_o}\right)^2, \quad 0 < \tau < \infty \tag{5.5}$$

We may identify t_0, τ_0 as the time coordinates of the present epoch.

(ii) Steady state model :

Here we will assume, without loss of generality $t = 0$, $\tau = 0$ to denote the present epoch. We then have

$$Q(t) = e^{Ht}, \quad H = \text{constant}, \quad \tau = \frac{1}{H}\left\{1 - e^{-Ht}\right\}$$

$$\Omega(\tau) = \frac{1}{1 - H\tau}, \quad -\infty < \tau < H^{-1}. \tag{5.6}$$

In the latter case note that the time axis on the τ scale ends at $\tau = H^{-1}$. This happens because there is an event horizon to the future of any fundamental observer. This fact will turn out to have very significant implications for quantum electrodynamics.

Exercises

1. Find the conformally flat metric of Brans-Dicke model for the general parameter A.

2. A Robertson-Walker model has an event horizon in the future. Show that if the metric of the model is expressed in the conformally flat form the conformal factor diverges at the horizon.

3. A light ray travelling into the future in the Einstein de Sitter universe has the starting wavelength of 4000 $\mathring{A}$. For the present epoch assume $t_0 = 12 \times 10^9$ years. After what period will the wavelength of light increase to 1 metre?

4. Show that in the expanding universes described here the redshift of a light source increases with distance and that the law for relatively nearby sources is linear. This is known as *Hubble's law*. Show also that the constant of proportionality, the Hubble constant is given by $\dot{Q}/Q$ evaluated at the present epoch.

LECTURE VI : RESPONSE OF THE EXPANDING UNIVERSE

A. Criterion for Absorption

With the inputs brought by cosmology it is now worth taking a second look at the absorber theory of radiation. There is, however, one subtle issue that Feynman (1963) had pointed out with regard to Hogarth's treatment of the problem that needs to be highlighted. For this, we go back to the general treatment of Lecture III.

As we pointed out, the condition for perfect absorption in the future demands [cf. Eq.(3.12)] that *as $r \to \infty$*

$$\sum_b \frac{1}{2}\left[F^{(b)ret} + F^{(b)adv}\right] \to 0 \quad \text{faster than} \quad 1/r. \tag{6.1}$$

In practice this is ensured by the absorptive part of the refractive index $n - ik$, i.e., by the parameter k. Hogarth had used the phenomenon of collisional damping to calculate k. Further, when he discussed the condition of perfect absorption in the past, he had used the same formula for k, *but with its sign reversed.* Feynman's criticism was that this sign reversal brought in thermodynamics that Hogarth was seeking to avoid. For, the phenomenon of collisional damping is a collective phenomenon that assumes the second law of thermodynamics and asymmetrical initial conditions. Thus, the claim that cosmology and not thermodynamics determined the unidirectionality of electromagnetic radiation was vitiated.

To get around Feynman's criticism, Hoyle and Narlikar (1963) proceeded in a different way: they used the radiation reaction on the charge to determine the damping parameter k. Their approach involved first choosing a particular combination of advanced and retarded solutions as the final solution and then testing it for self-consistency. Let us say that the pure retarded solution is to be so tested. Then, given that all charges interact finally through retarded waves, the radiation reaction is as given by Dirac [cf. Eq. (3.16)]. This reaction gives a force of damping which, in the non-relativistic limit leads to the following equation of motion for a typical absorber particle acted on by an external electric field **E**:

42

$$m\ddot{\mathbf{r}} = e\mathbf{E} + \frac{2e^2}{3}\dddot{\mathbf{r}} \qquad (6.2)$$

Here e is the charge and m the mass of the particle.

If we Fourier analyse with ω the angular frequency of a typical field component, then it is easy to see that the refractive index $n - ik$ of the absorber medium is given by the equation :

$$(n - ik)^2 = 1 - \frac{4\pi N e^2}{m\omega^2}\left\{1 + \frac{2ie^2\omega}{3m}\right\}^{-1}. \qquad (6.3)$$

Notice that in deriving the above equation and using it for calculating the imaginary part of the refractive index we have not gone beyond electrodynamics; certainly not to thermodynamics. Further, if we were testing the self-consistency of the advanced solutions we would likewise use Eq. (6.2) with the sign of the radiative reaction term reversed. This would change the Eq. (6.3) to

$$(n - ik)^2 = 1 - \frac{4\pi N e^2}{m\omega^2}\left\{1 - \frac{2ie^2\omega}{3m}\right\}^{-1}. \qquad (6.4)$$

This reversal of sign has no relationship to thermodynamics (as was the case with Hogarth's use of collisional damping), but follows logically from electrodynamics. Further, the presence of the imaginary part in the refractive index arising from the radiative reaction term tells us that we are not dealing with a pure scattering phenomenon.

B. Two Explicit Examples

The next step in the argument is from cosmology. Because a typical wave from the source undergoes a spectral shift while travelling into the past or the future, we have to take into account its changed frequency at the time of its interaction with the absorber particle.

To study this effect we will consider two explicit examples from Table I. These are illustrated in Fig. VI.1. First consider the Eistein deSitter model whose geometrical details were given in Eq. (5.5). We rewrite its line element in the Robertson–Walker form as

$$ds^2 = dt^2 - \left(\frac{t}{t_o}\right)^{4/3}[dr^2 + r^2(d\theta^2 + sin^2\theta d\phi^2)] \tag{6.5}$$

In manifestly conformally flat form this is

$$ds^2 = \left(\frac{\tau}{\tau_o}\right)^4[d\tau^2 - dr^2 - r^2(d\theta^2 + sin^2\theta d\phi^2)]. \tag{6.6}$$

Let us first test for the consistency of retarded solutions.

Suppose a general disturbance leaves the source at $r = 0$ at $t = t_0, \tau = \tau_o$, and travels along the future light cone. We consider a typical Fourier component of angular frequency ω of the electric field emanating from the accelerated source. At the source we have already adjusted the conformal factor of Eq. (6.6) to be unity. Thus the frequency ω measured on the t-scale is the same as that on the τ-scale. Since the electric field is conformally invariant it is convenient to work with the line element of Eq. (6.6) as we can take over the flat spacetime solution intact in these coordinates. However, as we found in the general treatment of Wheeler and Feynman, we need not go into specific details of the electric field but need only verify that it does indeed fall off faster than $1/r$ at large distances.

The flat spacetime expression tells us that without the interaction with the absorber, the field falls off as $1/r$. The absorber introduces a frequency dependent factor

$$\xi = exp\left(+ \int kdr \right) = exp(-I), \text{ say.} \tag{6.7}$$

of further damping. [Here $\omega(r)$ is the frequency of the wave at the radial coordinate r.] It is the asymptotic behaviour of this factor that decides whether the future absorber is perfect or not. If the integral in the exponent of Eq.(6.7) diverges, the absorption is complete; otherwise it is incomplete.

Our task therefore is reduced to computing the asymptotic form of the parameter k in the refractive index. To calculate this first note that although in the flat spacetime solution the frequency of the field does not change, the τ scale does not measure the proper frequency in the cosmological rest frame. Thus ω is not the proper frequency that the typical absorber particle at coordinate r interacts with. The proper frequency is measured on the t-scale and the value of the conformal factor at the absorber particle will determine it in terms of the constant frequency on the τ-scale. The proper frequency is therefore given by

$$\omega(r) = \omega\left[1 + \frac{r}{\tau_0}\right]^{-2}. \tag{6.8}$$

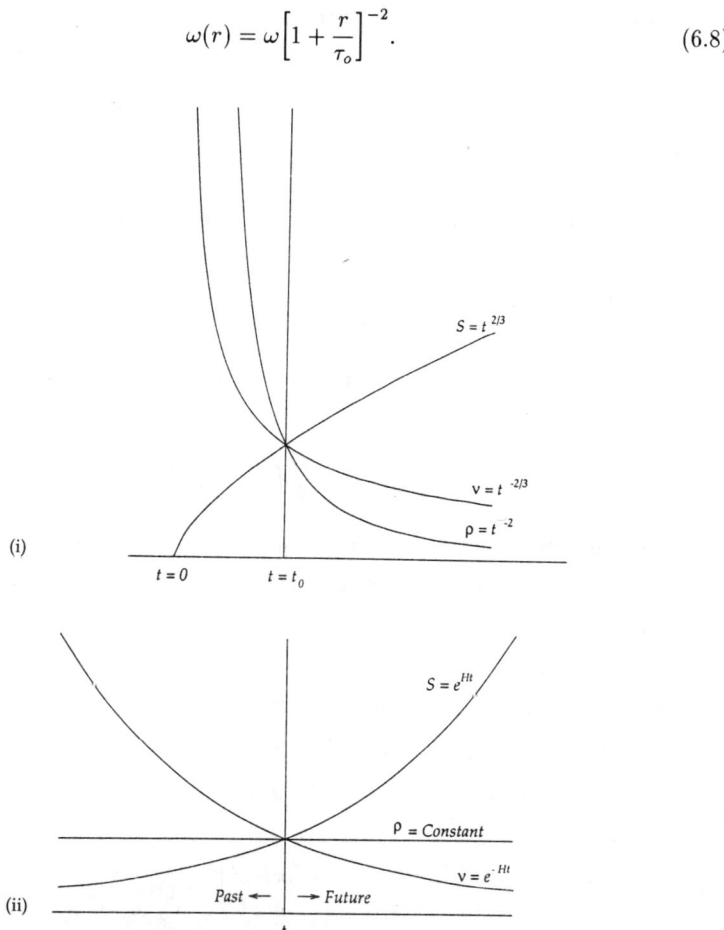

(i)

$t = 0$ $\quad$ $t = t_0$

$S = t^{2/3}$

$v = t^{-2/3}$

$\rho = t^{-2}$

(ii)

$S = e^{Ht}$

$\rho = Constant$

$v = e^{-Ht}$

Past $\leftarrow$ $\quad$ $\rightarrow$ Future

t_0

Fig. VI.1 In (i) we describe the epoch dependence of the scale factor, frequencies of signals into the past and the future and density in the Einstein de Sitter cosmology. These curves are to be compared and contrasted with those of (ii) for the steady state model. In each case ω denotes the angular frequency emanating from the source. All quantities are expressed in units of their values at the present epoch t_0.

A comparison with the Eq. (5.2) will tell us that the above is a restatement of the phenomenon of cosmological redshift. The absorber particle encounters a lower frequency than what was sent out by the source. The determination of k from the formula (6.3) requires us also to know the asymptotic behaviour of the number density N. A reference to Table I tells us that since the density ρ is proportional to N,

$$N(r) = N(0)\left[1 + \frac{r}{\tau_o}\right]^{-6}. \tag{6.9}$$

Using these formulae in Eq. (6.3) we find that the asymptotic value of k is given by

$$k \cong -\alpha\omega^2 \ , \qquad \alpha = \text{constant} \tag{6.10}$$

and the integral for absorption as given in Eq.(6.7) is

$$I \sim \int^\infty \alpha\omega^2 dr \sim \int^\infty \left(1 + \frac{r}{\tau_o}\right)^{-4} dr < \infty \tag{6.11}$$

Note that this integral converges, thus indicating that the absorption is imperfect. *It follows therefore that in the Einstein-deSitter cosmology, the retarded solution is not consistent.*

What about the advanced solution? We similarly consider the above formulae in the asymptotic limit of very large blueshifts. The situation at high energies is, however, not so clearcut. If we assume that interaction cross sections saturate as $\omega \to \infty$, then it can be shown [cf, Hoyle and Narlikar, 1963] that as $\omega \to \infty$,

$$1 - (n - ik)^2 = \frac{4\pi Ne^2}{m\omega^2}\left[1 + O\left(\frac{1}{\omega}\right)\right]. \tag{6.12}$$

Using the $N \propto \omega^3$ dependence and keeping in mind the fact that τ is bounded below at $\tau = 0$, the integral for absorption in the past becomes

$$I \propto \int_0^{\tau_o} \frac{dr}{\tau_o - r} = +\infty, \tag{6.13}$$

i.e., it diverges! Thus here we have the past absorber perfect and the future absorber imperfect, a situation leading to the advanced solutions being self-consistent.

We therefore have a cosmology that does distinguish between the past and the future absorbers, the main point made by Hogarth. The final outcome, however, is the opposite of what is found in real life. Let us now examine another case from Table I, the steady state model.

Using Eq. (5.6) it is easy to see that a retarded wave emitted at $t = 0$ by a source at $r = 0$ arrives at the absorber particle at the coordinate r at $\tau = r$ and therefore the frequency ω at the source is redshifted to $\omega(r) = \omega(1 - Hr)$. The number density of absorber particles per unit proper volume, however, remains constant at $N = N_0$, say. Again, evaluating the parameter k in the low frequency limit we finally get the absorption integral of Eq. (6.7) as

$$I = - \int_0^{H^{-1}} k dr \sim \int_0^{H^{-1}} \frac{dr}{1 - Hr} = \infty. \qquad (6.14)$$

This integral clearly diverges, thus ensuring perfect absorption in the future.

For the past absorber, a similar calculation gives the blueshifted frequency of the advanced wave at the absorber particle with coordinate r to be $\omega(r) = \omega(1 + Hr)$. Thus again we are dealing with high frequency waves at the asymptotic past infinity. However, the number density is still constant and hence from Eq. (6.12) the limiting value of the constant $-k$ is $\propto \omega(r)^{-3}$. So the absorption integral of Eq.(6.7) becomes

$$I = - \int^{\infty} k dr \propto \int^{\infty} (1 + Hr)^{-3} dr < \infty. \qquad (6.15)$$

This integral converges, indicating imperfect absorption.

This is another example of the past and future absorbers behaving differently. In this case, however, we do get the right answer, viz. that only the retarded solution is self-consistent.

The calculations with regard to the response of the past absorber as given above carry a caveat. The determination of the refractive index for very high energy waves cannot really be carried out entirely classically. Quantum effects cannot be ignored. Nevertheless, as stated earlier, if the quantum cross sections converge at high energy (as they must do) the conclusions drawn here will stand.

C. Responses of Different Universes

With regard to the models mentioned in Table I we find a variety of answers to the above type of calculation. The results are summarized in Table II below. For the cosmological models with the curvature parameter $= 1$ or -1 the calculation is more involved and was carried out by Roe (1969). Davies (1972) has also examined a whole class of cosmological models with somewhat different refractive indices. His conclusions in general are similar to that of Table II. He has, however, questioned whether the trapping of redshifted waves of frequencies below the plasma frequency in the future absorber of the steady state universe can be interpreted as 'absorption'. The point, however, is that whatever the physical process it will eventually ensure absorption of all waves of progressively decreasing frequencies as they travel through a future absorber of constant density and infinite extent. Thus the condition (3.12) is satisfied.

Davies (1973) has also pointed out that absorption will take place by macroscopic objects at all wavelengths; i.e., a galaxy will absorb photons at a rate proportional to the photon density ($\propto R^{-3}$) and hence if R increases slower than $t^{1/3}$ as in the Dirac model the time-integrated photon absorption diverges, giving perfect absorption. In the Dirac model with $G \propto 1/t$ the black hole radius tends to zero and so if all matter in galaxies etc. ultimately ends in black holes the universe would not be opeque. In this sense the Dirac model in Table II is a borderline case.

In some cases in the following table the models give both the advanced and retarded solutions as self-consistent. We call such a case ambiguous since there the cosmological time-asymmetry is not able to resolve the issue and we are not better off than Wheeler and Feynman working within the static universe.

Table II

Consistency of Advanced/ Retarded Solutions

Model	Future absorber	Past Absorber	Outcome
Einstein-deSitter	imperfect	perfect	advanced
Closed Friedmann	perfect	perfect	ambiguous
Open Friedmann	imperfect	perfect	advanced
Steady state	perfect	imperfect	retarded
Quasi-steady state	perfect	imperfect	retarded
Brans-Dicke	imperfect	perfect	advanced
Dirac	imperfect	perfect	advanced

Can we use collisional damping to settle these issues as Hogarth had attempted to do? In a self consistent picture the following must hold. If retarded solutions are to be justified, the future absorber must be perfect and the past absorber imperfect. Now a particular cosmological model may have a perfect future absorber using collisional damping; but how do we judge the efficacy of the past absorber? Being of thermodynamic origin the nature of the phenomenon along the past light cone cannot be determined unambiguously. Thus we cannot settle the issue without an *extra assumption* about the thermodynamic arrow of time. Hence, if we wish to work entirely within the framework of electrodynamics and cosmology we have to avoid the usage of collisional damping as the means of absorption. Once the consistency of retarded solutions is established, however, we can use the above process to compute any actual damping.

Where there is the correct (i.e., retarded) solution emerging clearly we are not only better off vis-a-vis Wheeler and Feynman but we are also better off compared to the classical field theory. For, in such models we are able to link the local electrodynamic time-asymmetry to the cosmological one and are thus able to demonstrate that the choice of retarded solutions is not *ad hoc* but forced by the universe. The analysis given here therefore answers the first of the questions raised at the end of Lecture I.

This gain is very important in rehabilitating action at a distance as a viable classical theory.

From this table it is clear that cosmologies with ongoing creation of matter deliver the right answer, because in their cases the future absorber has sufficient absorbing matter to be perfect while the past absorber is rarefied enough (for high frequency waves) to be imperfect. Thus, if a workable action at a distance theory is to be the decisive criterion, these theories have to be preferred to those others (like the big bang models of Friedmann) which give wrong or ambiguous answers. Considering, however, that this verdict on cosmology is the exact opposite to the current beliefs in the validity of big-bang models more has to be said in justification of action at a distance as the correct approach to electrodynamics.

In this context, the most important issues are raised in the questions 4 and 5 at the end of Lecture I. Can the action at a distance formulation be developed at the quantum level and does it throw more light on issues which trouble the field theory, e.g. the problem of self-action? We will review the progress on these fronts next.

Exercises

1. Discuss why Feynman objected to using the process of collisional damping as the fundamental process for settling the responses of past and future absorbers.

2. Show that the quasi-steady state cosmology has the correct responses from the past and future absorbers.

3. Comment on the relationship between the cosmological and electrodynamical time-asymmetry in expanding universes. Why is the action at a distance theory able to establish such a relationship whereas the field theory fails to provide one?

4. From standard textbooks in cosmology (see e.g. Narlikar 1993) evaluate the radiation background produced by a uniform distribution of identical sources in an expanding universe. How does the expansion of the universe help in limiting the magnitude of this background. Comment on the relationship between thermodynamics and cosmology with the help of this result.

PART II

QUANTUM ELECTRODYNAMICS - NON-RELATIVISTIC PROCESSES

LECTURE VII : THE PATH-INTEGRAL APPROACH TO
QUANTUM MECHANICS

A. Introduction

We have so far proceeded along classical lines. We have shown that the direct-particle approach to electromagnetism works at least as well as the Maxwell field approach in explaining all the classical phenomena of the interaction of charges, and that it links with cosmology in an interesting way. The choice of retarded potentials is not an *ad hoc* choice, but is dictated by the universe at large. Moreover, the unbounded motions of charges moving under the self-force do not arise in this theory.

Although success in the classical domain is necessary for any physical theory, it is not sufficient. Nature, as we understand it today, is quantum in character. In electrodynamics, quantum theory has unearthed a vast collection of phenomena outside the concepts of classical physics. These have been explained with remarkable success by the quantized version of Maxwell's theory, although conceptual and mathematical stumbling blocks have been encountered there too. Can the direct-particle theory do as well here, if not better?

At first sight, an attempt to extend the classical direct-particle theory to include quantum phenomena seems unlikely to succeed. In Maxwell's theory we have fields to quantize. The degrees of freedom of these fields result in packets of energy called "photons", which play such an important part in quantum electrodynamics. We have no analogous degrees of freedom in direct-particle fields. Thus photons do not appear to exist in the latter theory. The only degrees of freedom are those vested in the particles. Can we get all the conventional quantum electrodynamics by a first quantization alone? If not, the theory fails. If we can, however, the theory must be regarded as the superior theory, because it reproduces all observations under fewer degrees of freedom.

Other difficulties can be anticipated concerning particles and antiparticles. In classical theory all world lines are endless and timelike. In relativistic quantum electrodynamics, the world lines can go forward and backward in time. What happens to the "identity" of a world line under these circumstances? How does the rule of no self-interaction operate under such conditions?

These are some of the problems which arise when we undertake to quantize the Fokker theory. We shall proceed by stages in solving them, beginning with the simpler nonrelativistic picture and ending in Part III with fully relativistic interactions of electrons and positrons.

B. Path Amplitudes

Suppose a physical system has action S. This is defined in terms of "paths" Γ that the system can follow in coordinate space (see Fig. VII.1). Classical physics tells us that not all Γ are permissible. In general, there is a unique path Γ from a given point P_1 in coordinate space to a given point P_2. Writing Γ_0 for this path, Γ_0 is given by the principle of stationary action

$$\delta S = 0 \qquad \text{for } \Gamma = \Gamma_0. \tag{7.1}$$

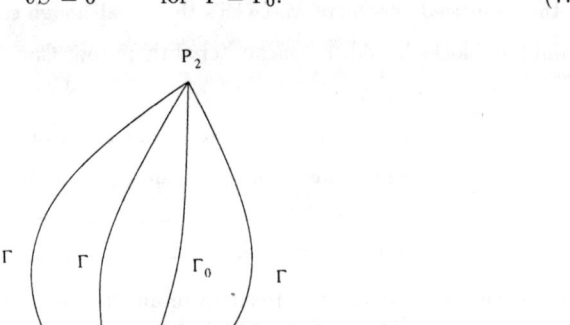

Fig. VII.1 The classical evolution of a physical system is along a unique path Γ_0 from the initial point P_1 to the final point P_2. In quantum mechanics there are several alternative paths Γ connecting P_1 to P_2, each with its own probability amplitude. The possibilities converge to Γ_0 as the action functional S becomes large compared to $\hbar$.

Earlier, we have used this mysterious prescription, and have noted its remarkable success in classical physics. In the thirties Dirac (1935) suggested the interesting idea, later developed quantitatively by Feynman, that (7.1) is the consequence of a more general principle operating in quantum mechanics. In quantum theory the system is permitted any of the paths Γ from P_1 to P_2, but each path has a definite probability amplitude proportional to

$$\exp\,(iS/\hbar), \tag{7.2}$$

where $\hbar$ is Planck's constant. All amplitudes add, giving a total amplitude for the system to go from P_1 to P_2. In the classical limit $\hbar \to 0$, and (7.2) oscillates wildly as we move from path to path – with the exception of Γ_0 where (7.1) holds. Paths in the neighborhood of Γ_0 make a significant contribution in this limit, whereas the contributions from other paths average out to zero. Hence the classical principle of stationary action.

Feynman carried these ideas further by introducing the concept of the path integral. He defined the nonrelativistic quantum-mechanical propagator $K(P_2; P_1)$ for the system to go from P_1 to P_2 by

$$K(P_2; P_1) = \sum_{\Gamma}(\text{constant}) \exp\,(iS/\hbar), \qquad t_2 > t_1,$$

$$\tag{7.3}$$

$$= 0, \qquad t_2 < t_1.$$

In (7.3) the action S is computed for each path Γ, and the sum is over all Γ from P_1 to P_2. The constant is a normalization constant. If, as is usual, the paths form a continuum, the sum is replaced by an integral

$$K(P_2; P_1) = \int \exp[iS(\Gamma)/\hbar]\mathcal{D}\Gamma, \qquad t_2 < t_1. \tag{7.4}$$

The integral is over the continuum of paths and is more complicated than the Riemann or the Lebesgue integral, which are summed over sets of points. Only limited progress has been made toward giving a rigorous mathematical foundation to this concept. Feynman was able, however, to obtain all required physical answers by various subtle devices. We shall draw heavily on these techniques. For details see Feynman and Hibbs (1965). The constant in (7.3) can be absorbed in the measure of $\mathcal{D}\Gamma$.

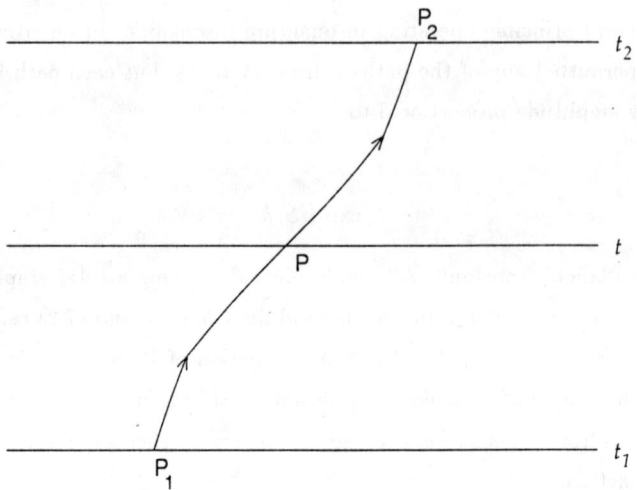

Fig. VII.2 Any path from P_1 to P_2 will cross the intervening time-plane at some point P. Thus, the total amplitude for the system to evolve from P_1 to P_2 will be made up by multiplying the amplitudes for the intervals (P_1, P) and (P, P_2) and adding for all P on the time plane.

Suppose P_1 represents the spacetime point (x_1, t_1) and P_2 the point (x_2, t_2) in the motion of a particle. Let $t_2 > t_1$. Any path Γ from P_1 to P_2 will pass through some intermediate point P having time coordinate t (see Fig. VII.2). Let the spatial position of P be x. Since the action functional is additive over paths, we can write

$$S(\Gamma_{P_2 P_1}) = S(\Gamma_{P_2 P}) + S(\Gamma_{P P_1}), \tag{7.5}$$

where the path $\Gamma_{P_2 P_1}$ from P_1 to P_2 is made up of the segment $\Gamma_{P P_1}$ from P_1 to P and the segment $\Gamma_{P_2 P}$ from P to P_2. Hence

$$\exp\,[iS(\Gamma_{P_2 P_1})/\hbar] = \exp\,[iS(\Gamma_{P_2 P})\hbar]\,\exp\,[iS(\Gamma_{P P_1})/\hbar]. \tag{7.6}$$

The sum over all paths from P_1 to P_2 can be obtained by summing all paths $\Gamma_{P P_1}$

and $\Gamma_{P_2 P}$ and integrating over the spatial coordinates of P. From (7.6) together with an appropriate measure for the path integral (7.4), we get

$$K(x_2, t_2; x_1, t_1) = \int K(x_2, t_2; x, t) K(x, t; x_1, t_1) d^3 x. \qquad (7.7)$$

Together, (7.4) and (7.7) suggest an alternative way of defining the path amplitude. Suppose we divide $\Gamma_{P_2 P_1}$ into a large number of small segments with intermediate points $Q_r (r = 1, \ldots, N-1)$. We can define $Q_0 = P_1$ and $Q_N = P_2$. Over each segment (Q_{r-1}, Q_r) we may imagine S to change very slowly. We define the amplitude for such a segment to be proportional to $K(Q_r; Q_{r-1})$. The amplitude along the entire path is then given by the product

$$\lim_{N \to \infty} \prod_{r=1}^{N} A_r K(Q_r; Q_{r-1}), \qquad (7.8)$$

where A_r is a constant of proportionality with the dimensions of spatial volume. Summing over all paths gives

$$\int \prod_{r=1}^{N} A_r K(Q_r; Q_{r-1}) \ \mathcal{D}\Gamma = \int \ldots \int K(P_2; Q_{N-1}) K(Q_{N-1}; Q_{N-2}) \ldots$$
$$\ldots K(Q_2; Q_1) K(Q_1; P_1) d^3 Q_{N-1} \ldots d^3 Q_1 = K(P_2; P_1) \qquad (7.9)$$

by (7.7), the integrations of $d^3 Q_r$ being over the time sections $t = t_r; r = 1, \ldots, N-1$.

The constants A_r are again absorbed in the measure of Γ. Because of (7.7), the definition (7.8) leads to the same function $K(P_2; P_1)$ as before.

Sometimes we know K but not S. Then (7.8) is useful to define a path amplitude. The original definition (7.2) is the more direct one, however. We shall use (7.8) in relativistic path-integral theory.

C. The Wavefunction and Schrödinger's Equation

Suppose that, instead of knowing that the particle is at (x_1, t_1), we only know the probability amplitude $\psi(x_1, t_1)$ for it to be at various spatial positions x_1 on the time section $t = t_1$. We then ask, what is the probability amplitude $\psi(x_2, t_2)$ of finding the particle at (x_2, t_2)? This is obtained from the weighted mean of $K(x_2, t_2; x_1, t_1)$,

$$\psi(x_2, t_2) = \int K(x_2, t_2; x_1, t_1)\psi(x_1, t_1)d^3x_1. \tag{7.10}$$

As $t_2 \to t_1, \psi(t_2) \to \psi(t_1)$. This implies

$$\lim_{t_2 \to t_1} K(x_2, t_2; x_1, t_1) = \delta_3(x_2 - x_1). \tag{7.11}$$

The function ψ is the usual Schrödinger wavefunction. Does it satisfy the Schrödinger equation? The answer is "yes", and we illustrate this by a simple example. For a particle moving in a potential field V, we have

$$S = \int \left(\frac{1}{2}m\dot{x}^2 - V\right)dt. \tag{7.12}$$

If we substitute this in (7.4) and use (7.10), we find that ψ satisfies the differential equation

$$-\frac{\hbar^2}{2m}\nabla_2^2\psi + V\psi = i\hbar\frac{\partial\psi}{\partial t_2}. \tag{7.13}$$

This is the one-dimensional Schrödinger equation. The same argument can readily be extended to obtain the three-dimensional Schrödinger equation.

In view of (7.11) and the fact that $K = 0$ for $t_2 < t_1$ (no propagation backward in time) (7.13) implies that K satisfies the equation

$$[(\partial/\partial t_2) - (i\hbar/2m)\nabla_2^2 + (i/\hbar)V] \ K(x_2, t_2; x_1, t_1)$$
$$= \delta_3(x_2 - x_1)\delta(t_2 - t_1). \tag{7.14}$$

D. Transition Probability

A practical question is, what will a system do under the action of specified forces? The quantum mechanical answer is in terms of probability that a system in a given initial state will make a transition to another given final state. The path integral description of this is briefly as follows.

Let the particle be in an initial state $\phi_i(x, t_1)$. We wish to determine the probability that it is in state $\phi_f(x, t_2)$ for $t_2 > t_1$. Using (7.10), the amplitude is given by

$$
\begin{aligned}
\langle \phi_f | \phi_i \rangle &= \int \int \phi_f^*(x_2, t_2) K(x_2, t_2; x_1, t_1) \phi_i(x_1, t_1) d^3x_2 d^3x_1 \\
&= \int \int \int \phi_f^*(x_2, t_2) \exp\left[iS(\Gamma_{21})/\hbar\right] \phi_i(x_1, t_1) \mathcal{D}\Gamma_{21} d^3x_2 d^3x_1,
\end{aligned}
\tag{7.15}
$$

where * denotes complex conjugate.

The transition probability from ϕ_i to ϕ_f is given by the square of the modulus of (7.15), i.e., by

$$
\begin{aligned}
P(\phi_i \rightarrow \phi_f) &= |\langle \phi_f | \phi_i \rangle|^2 \\
&= \int \int \int \int \int \int \phi_f^*(x_2, t_2) \exp\left[iS(\Gamma_{21})/\hbar\right] \phi_i(x_1, t_1) \\
&\quad \cdot \phi_i^*(x_1', t_1) \exp\left[-iS(\Gamma_{21}'/\hbar)\right] \phi_f(x_2', t_2) \\
&\quad \cdot \mathcal{D}\Gamma_{21} \mathcal{D}\Gamma_{21}' d^3x_1 d^3x_2 d^3x_1' d^3x_2'.
\end{aligned}
\tag{7.16}
$$

We shall often refer to Γ_{21}' as a conjugate path, implying that the action along Γ_{21}' is multiplied by $-i$ instead of $+i$.

Exercises

1. Relate the Feynman path integral prescription to the classical principle of stationary action in the limit $s \gg \hbar$.

2. Solve (7.14) for a free particle ($V = 0$) by taking Fourier transforms to show that

$$
K[x_2, t_2; x_1, t_1] = \sqrt{\frac{m}{2\pi i\hbar(t_2 - t_1)}} \exp\left[-\frac{im(x_2 - x_1)^2}{2\hbar(t_2 - t_1)}\right].
$$

3. Prove by direct integration that the transitive relation (7.7) holds for the free particle propagator.

LECTURE VIII : PERTURBATION THEORY AND
THE INFLUENCE FUNCTIONAL

A. A Power Series Expansion

Often the explicit evaluation of transition probability by the formula obtained at the end of the last lecture is not possible. It may, however, happen that the force disturbing a system is relatively modest. In which case one can use a power-series expansion of the exponential $\exp(-is/\hbar)$ in the hope that the answer can be evaluated term by term in a series. This gives us the familiar perturbation expansion that has been extensively used in quantum electrodynamics. We now consider this problem.

Suppose we know the behaviour of a system with action S_0. Let the system be disturbed by a potential $V(x,t)$ over the time interval (t_1, t_2). The action during this interval is therefore

$$S = S_0 - \int_{t_1}^{t_2} V(x,t)dt. \tag{8.1}$$

We now wish to determine $K_V(x_2, t_2; x_1, t_1)$ in terms of the unperturbed propagator $K_0(x_2, t_2; x_1, t_1)$. We have

$$K_0(x_2, t_2; x_1, t_1) = \int \exp\left[\frac{iS_0(\Gamma_{21})}{\hbar}\right] \mathcal{D}\Gamma_{21}, \tag{8.2}$$

$$K_V(x_2, t_2; x_1, t_1) = \int \exp\left[\frac{iS_0(\Gamma_{21})}{\hbar} - \frac{i}{\hbar}\int_{\Gamma_{21}} V dt\right] \mathcal{D}\Gamma_{21}. \tag{8.3}$$

Expanding the exponential involving the integral over V, then

$$K_V(x_2, t_2; x_1, t_1) = \int \exp\left[\frac{iS_0(\Gamma_{21})}{\hbar}\right] \cdot \left[1 - (i/\hbar)\int_{\Gamma_{21}} V dt \right.$$
$$\left. -(1/2\hbar^2)\left(\int_{\Gamma_{21}} V dt\right)^2 + \ldots\right] \mathcal{D}\Gamma_{21}. \tag{8.4}$$

The path integral for the unity term in this expansion gives K_0. We consider the next term

$$-\int \exp\left[\frac{iS_0(\Gamma_{21})}{\hbar}\right]\cdot\frac{i}{\hbar}\int_{\Gamma_{21}} V\,dt\,\mathcal{D}\Gamma_{21}. \tag{8.5}$$

The integral $\int_{\Gamma_{21}} V\,dt$ is over a specific path Γ_{21}, given by a function $x(t)$. Suppose we take a particular instant t of time and take all paths which pass through $x(t)$ on their way from P_1 to P_2. If we sum over these paths alone, we will get, as in the analysis leading to (7.7), the product

$$K_0(x_2, t_2; x, t)K_0(x, t; x_1, t_1). \tag{8.6}$$

If the integral over V were absent, we would simply have got (7.7). But now we have to weight (8.6) with $-iV(x,t)/\hbar$. Then we have to sum over all x, to include all paths from 1 to 2. Finally we perform the time-integral to get the entire contribution :

$$-\frac{i}{\hbar}\int_{t_1}^{t_2} K_0(x_2, t_2; x, t)V(x,t)K_0(x, t; x_1, t_1)d^3x\,dt. \tag{8.7}$$

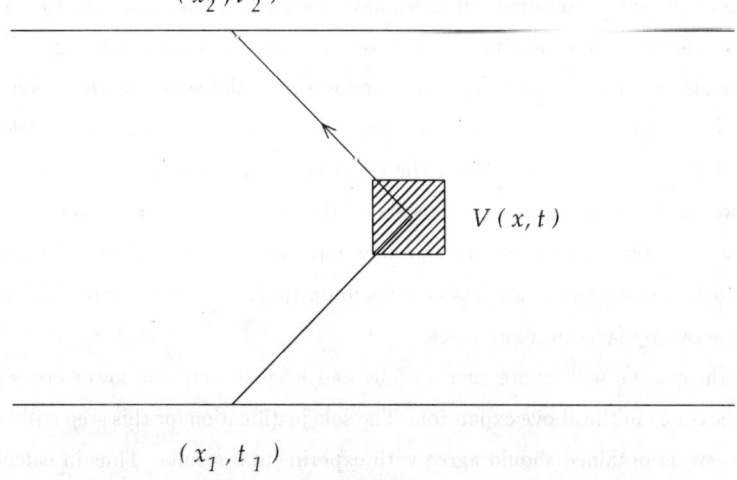

Fig. VIII.1 In the perturbation expansion the effect of an external potential V is built in through successive scatterings of the free particle propagator. Above we have the simplest example of single scattering by V.

The physical meaning of this operation is illustrated in Fig. VIII.1. If we consider a given t and x, we imagine the system to proceed undisturbed from (x_1, t_1) to (x, t). Then it is scattered by $V(x, t)$, after which it proceeds undisturbed from (x, t) to (x_2, t_2). This scattering could occur anywhere within the spacetime slab $t_1 \leq t \leq t_2$. Hence the integration in (8.7). The four-dimensional volume element, $d^3 x \, dt = d\tau$, say, can actually be taken over the whole of spacetime, since one of the K_0 functions involves propagation backward in time. The integrand is therefore zero whenever the point x, t, falls outside the slab $t_1 \leq t \leq t_2$.

Higher-order terms in the expansion give multiple scattering processes. Thus (8.3) is the closed form of the usual infinite perturbation series :

$$K_V(x_2, t_2; x_1, t_1) = K_0(x_2, t_2; x_1, t_1) - \frac{i}{\hbar} \int K_0(x_2, t_2; x, t) V(x, t) K_0(x, t; x_1, t_1) d\tau$$

$$+ \left(\frac{i}{\hbar}\right)^2 \int \int K_0(x_2, t_2; x_3, t_3) V(x_3, t_3) K_0(x_3, t_3; x_4, t_4) V(x_4, t_4). K_0(x_4, t_4; x_1, t_1) d\tau_3 d\tau_4 + \ldots$$
$$(8.8)$$

In the above expansion many liberties have been taken with mathematical rigour. Thus, we do not know, *a-priori*, if any of the above integrals converge. Even if all integrals in all terms converge, we cannot say if the series in (8.8) converges. In practice, if we consider interactions between electrons, as charged particles, the potential V has an e^2 term, e being the electron charge. Thus, with e restored, the expansion in (8.8) has increasing powers of the fine structure constant $(e^2/\hbar c) \sim 1/137$. Since this is a smallish fraction one may *assume* that higher order terms in the perturbation expansion are less and less important. This assumption, is, however, not borne out by facts in many cases.

Nevertheless, we will ignore such pitfalls and keep to only the lower order terms (first or second) in the above expansion. The sole justification for this step will be that the answers so obtained should agree with experimental results. Thus in calculating the transition probabilities via Eq.(7.16) we may use the above approximation.

B. Transition Element

Classically we are used to continuous changes of dynamical variables. In quantum mechanics, transitions are in general discontinuous. Can we give meaning to "velocity" or "acceleration" in discontinuous transitions? In the path-integral formulation, we can indeed give a meaning to such terms, by means of the concept of transition elements. Suppose ϕ_i and ϕ_f are the initial and final states of a system described by action S. We have already seen that the probability amplitude can be defined by the path integral

$$\langle \phi_f | \phi_i \rangle = \int \int \int \phi_f^* \exp\,(iS/\hbar)\phi_i \mathcal{D}\Gamma d^3 x_1 d^3 x_2. \tag{8.9}$$

Suppose we now have a functional $F[\Gamma]$ of a path Γ. Then the transition element of F is given by

$$\langle \phi_f | F | \phi_i \rangle = \int \int \int \phi_f^* \exp\left[\frac{iS(\Gamma)}{\hbar}\right] F[\Gamma]\phi_i \mathcal{D}\Gamma d^3 x_1 d^3 x_2. \tag{8.10}$$

This definition means that $\langle \phi_f | F | \phi_i \rangle$ is a certain kind of average over all paths suitably weighted by the initial and final wavefunctions. Classically, we could have calculated $F[\Gamma]$ exactly, since we know that $\Gamma = \Gamma_0$ is the unique solution. In quantum mechanics, we cannot make such a definitive statement.

Equation (8.10) is easy to apply to simple cases. For instance, for a free particle of mass m, the transition element of velocity is given by

$$\langle \phi_f | \dot{x} | \phi_i \rangle = \int -\frac{i\hbar}{m}\phi_f^* \nabla_x \phi_i d^3 x. \tag{8.11}$$

It can be easily verified in this example that a transition element need not be real even if the original dynamical variable is real.

C. Influence Functional

We come now to the concept most useful for the quantum development of direct-particle theories. Suppose we have one quantum-mechanical system in interaction with another, and suppose we are interested only in the detailed behaviour of the first system regardless of what happens to the second. We may speak of the second system as the "external environment" of the first. To fix ideas, let the first system be described by a coordinate q and the second by Q. The combined action for the two systems is taken to be of the form

$$S = S_0[q(t)] + S_E[Q(t)] + S_I[q(t), Q(t)], \qquad (8.12)$$

where S_0 represent sthe action of the first system alone, S_E is the action of the second system alone, and S_I represents the interaction of the two systems. Let $\phi_i(q_i, t_i)$ be the initial state and $\phi_f(q_f, t_f)$ the final state of the first system. Using (7.7), the probability of transition $\phi_i \to \phi_f$ is given by

$$P(\phi_i \to \phi_f) = \int \int \int \int \int \int \phi_f^*(q_f, t_f) \phi_i^*(q_i', t_i) \phi_i(q_i, t_i) \phi_f(q_f', t_f)$$

$$\qquad (8.13)$$

$$.\exp\left(\frac{i}{\hbar}\{S_0[q(t)] - S_0[q'(t)]\}\right) F[q, q'] \mathcal{D}q \mathcal{D}q' dq_f dq_f' dq_i dq_i',$$

where

$$F[q(t), q'(t)] = \sum_f \langle \psi_f | \exp\left\{\frac{i}{\hbar} S_I[q(t),\ Q(t)]\right\} | \psi_i \rangle$$

$$\qquad (8.14)$$

$$.\langle \psi_f | \exp\left\{\frac{i}{\hbar} S_I[q'(t), Q'(t)]\right\} | \psi_i \rangle^*.$$

Here ψ_i is the initial state of the second system. We have summed over all final states ψ_f, since we are not interested in how the environment is affected in detail.

$F[q(t), q'(t)]$ is called the *influence functional* and represents the "force" exerted by the environment on the first system.

Feynman and Hibbs (op. cit.) discuss influence functionals and their applications in some detail; so we shall not discuss them further. We expect the universe to act as the external environment in the quantum version of the Wheeler-Feynman theory. It is our aim to determine the quantum analogue of the response of the universe that earlier we calculated classically. This response will appear in the form of an influence functional.

Exercises

1. Show that (8.8) is equivalent to the standard perturbation expansion of Schrödinger's equation.

2. Draw an analogy between the external force in classical mechanics and the influence functional in quantum mechanics.

LECTURE IX : ABSORPTION AND STIMULATED EMISSION

A. Statement of the Problem

While a first quantisation of particles suffices to determine transition probabilities with respect to a prescribed potential function, as in Lectures VII and VIII, transitions in a radiation field are usually considered to require a quantisation of the field itself – i.e. a resolution into photons rather than classical wave theory. Then to match the quantisation of the field it is usual to adopt a second quantisation of the particles. However, we shall show here that none of this is necessary, all the standard radiation results being obtainable without any more quantisation than that given in A being needed.

In a simplified notation the probability $P(m \rightarrow n)$ defined by (7.16), of transition in the time interval $0 \leq t \leq T$ from the state m to an orthogonal state n in a specified (unquantised) external field is given by

$$P(m \rightarrow n) = \int \int \int \int \phi_n^*(\mathbf{a}_f).\phi_n(\mathbf{a}_f') J \phi_m(\mathbf{a}_i) \phi_m^*(\mathbf{a}_i') d^3\mathbf{a}_f d^3\mathbf{a}_f' d^3\mathbf{a}_i d^3\mathbf{a}_i', \qquad (9.1)$$

where ϕ_m, ϕ_n are the wave functions for the states, and J is the double path integral

$$J = \int \int \exp\left[\frac{i}{\hbar}\{S[\mathbf{a}(t)] - S[\mathbf{a}'(t)]\}\right] \mathcal{D}^3\mathbf{a}(t)\mathcal{D}^3\mathbf{a}'(t). \qquad (9.2)$$

Here the path $\mathbf{a}(t)$ "begins" at $\mathbf{r} = \mathbf{a}_i$, $t = 0$ and "ends" at $\mathbf{r} = \mathbf{a}_f$, $t = T$, while the path $\mathbf{a}'(t)$ begins at $\mathbf{r} = \mathbf{a}_i'$, $t = 0$ and ends at $\mathbf{r} = \mathbf{a}_f'$, $t = T$. The action $S[\mathbf{a}(t)]$ is a functional of the path $\mathbf{a}(t)$ and is defined by

$$S[\mathbf{a}(t)] = \int_0^T L(\mathbf{a}, t)dt, \qquad (9.3)$$

and similarly for $S[\mathbf{a}'(t)]$.

We are concerned with an electronic transition and with a situation in which

$$L = \frac{1}{2}m\dot{\mathbf{a}}^2 + eV(\mathbf{a}, t) - e\dot{\mathbf{a}}.\mathbf{A}(\mathbf{a}, t), \qquad (9.4)$$

66

where e, m are the electronic charge and mass, the velocity of light is taken as unity, the nonrelativistic kinetic energy is used, and $V(\mathbf{a}, t)$, $\mathbf{A}(\mathbf{a}, t)$ are the scalar and vector potentials of the specific field. In a simple one-electron atomic problem the external field is made up of the electrostatic field within the atom together with the field incident on the atom. Provided the latter has no electrostatic component it yields the vector potential $\mathbf{A}$, which in the Coulomb guage satisfies

$$div\,\mathbf{A} = 0. \tag{9.5}$$

The external field $\mathbf{A}$ is then wholly transverse. With this division of V and $\mathbf{A}$, write

$$S[\mathbf{a}(t)] = S_0[\mathbf{a}(t)] - e \int_0^T \dot{\mathbf{a}}.\mathbf{A}\,dt, \tag{9.6}$$

so that $S_0(\mathbf{a}(t))$ is the action for the atomic field alone.

Thus, our problem is to describe the transition probability of an atomic electron in stationary state m with energy E_m going to stationary state n with energy E_n when acted on by a external direct particle field of other charges. That the field is described by a vector potential $\mathbf{A}$ only implies that it has arisen from the acceleration of these charges, their net Coulomb field being negligible.

B. The Transition Probability

Next we regard $\mathbf{A}$ as small enough for the series

$$\exp\left[-\frac{ie}{\hbar} \int_0^T \dot{\mathbf{a}}.\mathbf{A}\,dt \right] = 1 - \frac{ie}{\hbar} \int_0^T \dot{\mathbf{a}}.\mathbf{A}\,dt + \ldots \tag{9.7}$$

to be rapidly convergent. The dominant term in (9.1) then involves the product of the first-order term in (9.6) with the first-order term in the complex conjugate of (9.6). After a reduction that is straightforward, except perhaps for the calculation of the transition element of the velocity (Feynman and Hibbs, p. 184) one obtains

$$P(m \to n) = \frac{e^2}{\hbar^2}\left(\frac{E_n - E_m}{\hbar} \right)^2 \left| \int_0^T \exp\frac{i(E_n - E_m)t}{\hbar} \int \phi_n^*(\mathbf{a})\mathbf{a}.\mathbf{A}\phi_m(\mathbf{a})d^3\mathbf{a}dt \right|^2. \tag{9.8}$$

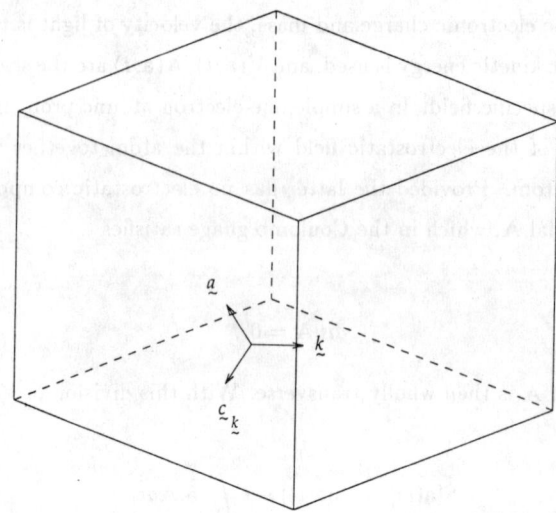

Fig. IX.1]The atom with typical alignment vector **a** is shown inside a large cubical box. The vectors **k** and **c$_k$** are the wavelength and amplitude vectors of a typical Fourier component of the direct particle field acting on the atom, the two vectors being orthogonal.

To express this result in a more familiar form expand the incident field in a Fourier series. For this step consider the atom as situated inside a cube with side of unit length, the latter being chosen to be very large compared with any wavelength of importance in the transition problem. Fig. IX.1 describes the geometry. Write

$$\mathbf{A}(\mathbf{a}, t) = \sqrt{4\pi} \sum_{\mathbf{k}} \{\mathbf{c}_k \exp\,[i(\mathbf{k}.\mathbf{a} - kt)] + \mathbf{c}_\mathbf{k}^* \exp\,[-i(\mathbf{k}.\mathbf{a} - kt)]\}, \qquad (9.9)$$

noting that $\mathbf{k} = 2\pi(n_1, n_2, n_3)$ where n_1, n_2, n_3 are integers, and that $\mathbf{k}.\mathbf{c}_k = 0$ because of the Coulomb gauge. Inserting (9.9) in (9.8) leads to a double sum, $\sum_\mathbf{k} \sum_\mathbf{k}$ say. Waves with $\mathbf{k} \neq \mathbf{k}'$ are regarded as being in random phase with respect to each other. After averaging $\sum_\mathbf{k} \sum_\mathbf{k}$ thus reduces to $\sum_\mathbf{k}$. Moreover, T is large enough so that even for allowed transitions there are a very large number of atomic oscillations in time T thus justifying the approximation

$$\left| \int_0^T \exp\,\left[\frac{i}{\hbar}(E_n - E_m \pm \hbar k)t\right]dt \right|^2 \simeq 2\pi T \delta\left(k \pm \frac{E_n - E_m}{\hbar}\right). \qquad (9.10)$$

When $E_n > E_m$ the important terms involve the minus sign in the delta-function, and (9.8) leads to

$$P(m \to n) = (4\pi e^2/\hbar)^2 \left(\frac{E_n - E_m}{\hbar}\right)^2 \sum_{\mathbf{k}} \left| \mathbf{c_k} \cdot \int \phi_n^*(\mathbf{a}) \mathbf{a} e^{i\mathbf{k} \cdot \mathbf{a}} \phi_m(\mathbf{a}) d^3\mathbf{a} \right|^2$$

$$\times \left| \int_0^T \exp\left[\frac{i}{\hbar}(E_n - E_m - \hbar k)t\right] dt \right|^2,$$

(9.11)

where we have deferred using (9.10) until after the sum has been converted to an integral. This is the case of absorption.

When $E_m > E_n$ we have the case of stimulated emission, and the corresponding result is

$$P(m \to n) = (4\pi e^2/\hbar^2) \left(\frac{E_m - E_n}{\hbar}\right)^2 \sum_{\mathbf{k}} \left| \mathbf{c_k^*} \cdot \int \phi_n^*(\mathbf{a}) \mathbf{a} e^{-i\mathbf{k} \cdot \mathbf{a}} \phi_m(\mathbf{a}) d^3\mathbf{a} \right|^2$$

$$\times \left| \int_0^T \exp\left[\frac{i}{\hbar}(E_n - E_m + \hbar k)t\right] dt \right|^2.$$

(9.12)

It is not hard to see that (9.12) is the same as (9.11). Although $\mathbf{c_k^*} e^{-i\mathbf{k} \cdot \mathbf{a}}$ has replaced $\mathbf{c_k} e^{i\mathbf{k} \cdot \mathbf{a}}$ the roles of the states m, n have been switched; ϕ_m is the state of lower energy in (9.11) whereas ϕ_n is the wavefunction of the lower energy state in (9.12). This confirms the introductory remark that the transition probability for stimulated emission in an external field is equal to that for absorption.

C. Relationship to Field Intensity

It is usual to express the transition probability in terms of the intensity of the applied field. In order to do this it is necessary either to approximate (9.11) or to average the transition probability with respect to the orientation of the atom. We adopt the latter procedure. Define

$$\mathbf{a}_{mn}(\mathbf{k}) = \int \phi_n^*(\mathbf{a}) \mathbf{a} e^{i\mathbf{k} \cdot \mathbf{a}} \phi_m(\mathbf{a}) d^3\mathbf{a},$$

(9.13)

and let $\alpha_{\mathbf{k}}^{(1)}, \alpha_{\mathbf{k}}^{(2)}$ be unit vectors which together with $\mathbf{k}/k$ form an orthogonal triad. Because of $\mathbf{c_k} \cdot \mathbf{k} = 0$

$$c_{\mathbf{k}} = \sum_{j=1,2} [c_{\mathbf{k}}.a_{\mathbf{k}}^{(j)}]\alpha_{\mathbf{k}}^{(j)}, \qquad (9.14)$$

and

$$|c_{\mathbf{k}}.a_{mn}(\mathbf{k})|^2 = \left| \sum_{j=1,2} (c_{\mathbf{k}}.\alpha_{\mathbf{k}}^{(j)}) \int \phi_n^*(\mathbf{a})\alpha_{\mathbf{k}}^{(j)}.ae^{i\mathbf{k}.\mathbf{a}}\phi_m(\mathbf{a})d^3\mathbf{a} \right|^2. \qquad (9.15)$$

We have to average (9.15) with respect to orientation.

It is not hard to show that the sum of

$$\int \phi_n^*(\mathbf{a})\alpha_{\mathbf{k}}^{(1)}.ae^{i\mathbf{k}.\mathbf{a}}\phi_m(\mathbf{a})d^3\mathbf{a}. \int \phi_n(\mathbf{a})\alpha_{\mathbf{k}}^{(2)}.ae^{-i\mathbf{k}.\mathbf{a}}\phi_m^*(\mathbf{a})d^3\mathbf{a}$$

and its complex conjugate averages to zero, and that (9.15) is proportional to $|c_{\mathbf{k}}|^2$ and can be written in the form $\frac{1}{3}|c_{\mathbf{k}}|^2|a_{mn}(k)|^2$, where $|a_{mn}(k)|^2$ depends on the magnitude but not the direction of $\mathbf{k}$, and is equal to $|a_{mn}|^2$ when the factor $\exp(i\mathbf{k}.\mathbf{a})$ in (9.15) is approximated by unity. Hence the average value of (9.11) is

$$\bar{P} = \frac{4\pi e^2}{\hbar^2}\left(\frac{E_n - E_m}{\hbar}\right)^2 \sum_{\mathbf{k}} |c_{\mathbf{k}}|^2.|a_{mn}(k)|^2. \left| \int_0^T \exp\left[\frac{i}{\hbar}(E_m - E_n + \hbar k)t\right]dt \right|^2. \qquad (9.16)$$

There are $d^3\mathbf{k}/(2\pi)^3$ terms of this series in the element $d^3\mathbf{k}$ of $\mathbf{k}$-space. Defining $\overline{|c_{\mathbf{k}}|^2}$ by

$$\overline{|c_{\mathbf{k}}|^2}\frac{d^3\mathbf{k}}{(2\pi)^2} = \sum_{d^3\mathbf{k}} |c_{\mathbf{k}}|^2,$$

we now write (9.16) in the integral form

$$\bar{P} = \frac{e^2}{6\pi^2\hbar^2}\left(\frac{E_n - E_m}{\hbar}\right)^2 \int \overline{|c_{\mathbf{k}}|^2}.|a_{mn}(k)|^2. \left| \int_0^T \exp\left[\frac{i}{\hbar}(E_n - E_m - \hbar k)t\right]dt \right|^2.d^3\mathbf{k} \qquad (9.17)$$

To define the intensity $I(\mathbf{k})$ per unit solid angle we note first that

$$\frac{1}{4\pi}\mathbf{E} \times \mathbf{H} = 2\sum_{\mathbf{k}} |c_{\mathbf{k}}|^2 k\mathbf{k} \doteq \frac{1}{4\pi^3}\int \overline{|c_{\mathbf{k}}|^2}k\mathbf{k}d^3\mathbf{k} \qquad (9.18)$$

and that the contribution of solid angle $d\Omega$ to (9.18) is

$$\frac{d\Omega}{4\pi^3} \cdot \frac{\mathbf{k}}{k} \int \overline{|c_\mathbf{k}|^2} k^4 dk, \qquad (9.19)$$

where $\mathbf{k}$ lies in $d\Omega$. The intensity $I(\mathbf{k})$ is now defined by equating (9.19) to

$$d\Omega . \frac{\mathbf{k}}{k} \int I(\mathbf{k}) dk.$$

For this definition to hold irrespective of $\overline{|c_\mathbf{k}|^2}$ we must have

$$I(\mathbf{k}) = \frac{1}{4\pi^3} k^4 \overline{|c_\mathbf{k}|^2}. \qquad (9.20)$$

Eliminating $\overline{|c_\mathbf{k}|^2}$ between (9.16) and (9.19), and using (9.9), one easily obtains

$$\bar{P}(m \to n) = \frac{4\pi^2 e^2}{3\hbar^2} \int d\Omega \int_0^\infty I(\mathbf{k}) |a_{mn}(k)|^2 \delta\left(k - \frac{|E_n - E_m|}{\hbar}\right) dk \qquad (9.21)$$

for the transition per unit time. The result (9.21) applies both to absorption and stimulated emission, and is the usual relation between the intensity and the transition probability. We can introduce separate intensities $I^{(j)}(\mathbf{k})$ for the "polarisation" directions $\alpha_\mathbf{k}^{(j)}$ by defining

$$I^{(j)}(\mathbf{k}) = \frac{k^4}{4\pi^3} \overline{|c_\mathbf{k} . \alpha_\mathbf{k}^{(j)}|^2}; \qquad j = 1, 2. \qquad (9.22)$$

Evidently

$$I(\mathbf{k}) = \sum_{j=1,2} I^{(j)}(\mathbf{k}). \qquad (9.23)$$

D. Opacity

We end the present Lecture by introducing the concept of opacity. Suppose there are $n(k)dk$ atoms per unit volume in state m, where $E_m < E_n$ and $(E_n - E_m)/\hbar$ lies between k and $k + dk$. Define a mean value of $|a_{mn}(k)|^2$ by

$$V\overline{|\mathbf{a}_{mn}(k)|^2}.n(k)dk = \sum_V |\mathbf{a}_{mn}(k)|^2, \tag{9.24}$$

the summation being taken through the small volume V for all atoms with $(E_n - E_m)/\hbar$ between k and $k + dk$. Both $\overline{|\mathbf{a}_{mn}(k)|^2}$ and $n(k)$ can be functions of position as well as of k.

Suppose radiation of frequency k travels along a three-dimensional spatial path Γ connecting two points P_1 and P_2. Then the opacity difference between these points is $\int_\Gamma d\tau$, where $d\tau$ is the difference for an element of the path. To obtain $d\tau$, let s be three-dimensional length along Γ and let $\mathbf{u}(s)$ be the unit tangent vector at s. The opacity differential $d\tau(k)/ds$ for frequency k is defined in terms of absorption by the equation

$$I(k\mathbf{u})d\Omega dk\frac{d\tau(k)}{ds} = \frac{4\pi^2 e^2}{3\hbar^2}d\Omega I(k\mathbf{u})\overline{|\mathbf{a}_{mn}(k)|^2}\hbar kn(k)dk, \tag{9.25}$$

and the opacity difference is

$$\int_{P_1}^{P_2} d\tau(k) = \frac{4\pi^2 e^2}{3\hbar}.k\int_{P_1}^{P_2} \overline{|\mathbf{a}_{mn}(k)|^2}n(k)ds. \tag{9.26}$$

Radiation travelling from P_1 and P_2 is reduced in intensity by the factor $\exp[-\int_{P_1}^{P_2} d\tau]$, it being supposed that enough atoms are involved for the absorption probability to be averaged. It may be noted that the factor $\hbar k$ appears on the right hand side of (9.25) because of $E_n - E_m \cong \hbar k$, not because of field quantisation.

Exercises

1. In field theory one can make arbitrary gauge transformations; but a direct particle field comes with a fixed gauge. How than can we justify the choice of gauge in Eq. (9.5)?

2. Derive the result of Eq. (9.10).

3. Deduce Eq. (9.18).

4. Relate the expression for opacity $\tau(k)$ in Eq. (9.25) to the classical concepts of absorption in Part I.

LECTURE X : SPONTANEOUS EMISSION

A. Introduction

We come now to a crux in the discussion of this book, to a problem which unless it can be solved would end the entire development. Whereas in Maxwell's theory the field has independent degrees of freedom that, after quantisation, generate spontaneous emission from the excited states of atoms, here we have no such resource. How then is spontaneous emission to be understood and calculated?

The solution to this problem will be sought along the following lines. In the quantum mechanical analogue of the problem of accelerated charge discussed in Lecture III, we will consider an atomic electron a in a specified stationary state i with energy E_i. Let us denote the state by a wavefunction $\psi_i(\mathbf{a}, t)$. If $K(\mathbf{a}_2, t_2, \mathbf{a}_1, t_1)$ is the propagator for the atomic problem, then wavefunction at a later time is given by

$$\begin{aligned} \psi(\mathbf{a}_2, t_2) &= \int K(\mathbf{a}_2, t_2; \mathbf{a}_1, t_1)\psi_i(\mathbf{a}_1, t_1)d^3\mathbf{a}_1 \\ &= \psi_i(\mathbf{a}_2, t_2), \end{aligned} \tag{10.1}$$

as indeed expected for a 'stationary' state.

However, a closer examination tells us that the assumption of stationarity is valid for an isolated atom. For an atom in a well-filled universe, the atomic electron is in direct contact with the past and future absorbers. Hence, any path Γ from $(\mathbf{a}_1, t_1)$ to $(\mathbf{a}_2, t_2)$ will, in general describe an accelerated electron which evokes an instantaneous response of the universe. Thus, the propagator K used in (10.1) is incomplete and hence incorrect, as it does not include this response. The complete propagator should take into account the interaction of all geometrically permissible paths Γ with the absorbers at large distances. This calculation is naturally quite involved; but we will go through it step by step to illustrate how the direct particle theory works in the quantum mechanical framework.

B. The Influence of the Future Absorber

To fix ideas, let us consider the motion of particle a in the time-interval $0 \leq t \leq T$

(see Fig. X.1), and denote the displacement of a by $\mathbf{a}(t)$. Let b be a typical absorber particle whose world line is intersected in intervals $\mathbf{\Delta}_-$ and $\mathbf{\Delta}_+$, respectively, by the past and future light cones from the initial point $[\mathbf{a}(0), 0]$ and the final point $[\mathbf{a}(T), T]$ on the path $\mathbf{a}(t)$ of a. From what has been said above, the induced transitions of a arise from its interaction with the retarded field of b, i.e., from the portion $\mathbf{\Delta}_-$ of the world line of b. The action governing induced transitions is therefore

$$-e_a \int_0^T \mathbf{A}_{\text{ret}}^{(b)}(\mathbf{a}).\dot{\mathbf{a}}\,dt, \tag{10.2}$$

where $\mathbf{A}_{\text{ret}}^{(b)}$ is the full retarded 3-potential from b. To calculate spontaneous transitions, we need, on the other hand, the transitions of b induced by the full retarded field of a. Further, in analogy with the classical calculation of Lecture III, we will need to calculate the advanced effect of b's motion (induced by a) back on a. (In what follows, even if all charges are of the same magnitude, being those of electrons, we will retain suffixes $e_a, e_b, \ldots$ to identify them.) The action governing the former is

$$-e_b \int_{\mathbf{\Delta}_+} \mathbf{A}_{\text{ret}}^{(a)}(\mathbf{b}).\dot{\mathbf{b}}\,dt, \tag{10.3}$$

with a similar notation. As for the latter, the motion of b will generate an influence on a. Following the classical case, we will look for a self-consistent cycle of argument in which the net field is the retarded one. Throughout this calculation, we shall work in a conformally flat cosmological model with a perfect future absorber and an imperfect past absorber. Thus the response comes only from the future.

Our eventual aim is to calculate the influence functional governing the motion of a and arising from the *whole universe*. Since we may assume the different absorber particles to act independently, this influence functional has the form

$$F[\mathbf{a}(t), \mathbf{a}'(t)] = \prod_{b \neq a} F^{(b)}[\mathbf{a}(t), \mathbf{a}'(t)], \tag{10.4}$$

where $F[\mathbf{a}(t), \mathbf{a}'(t)]$ is the influence functional exerted by a typical particle b.

To calculate $F^{(b)}[\mathbf{a}(t), \mathbf{a}'(t)]$ we consider all transitions produced by (10.3) in the absorber particle b. If $\psi_i(\mathbf{b})$ is the initial wavefunction and $\psi_f(\mathbf{b})$ any final wavefunction of b, we have from (8.14)

$$F^{(b)}[\mathbf{a}(t), \mathbf{a}'(t)] = \sum_f \int \int \int \int \psi_f^*(\mathbf{b}_f)\psi_f(\mathbf{b}_f')J^{(b)}\psi_i(\mathbf{b}_i)\psi_i^*(\mathbf{b}_i')d^3\mathbf{b}_id^3\mathbf{b}_fd^3\mathbf{b}_i'd^3\mathbf{b}_f',$$

$$(10.5)$$

where

$$J^{(b)} = \int \int \exp \frac{i}{\hbar}\{S_E[\mathbf{b}(t)] - S_E[\mathbf{b}'(t)] \; + S_I[\mathbf{a}(t), \mathbf{b}(t)]$$

$$-S_I[\mathbf{a}'(t), \mathbf{b}'(t)]\}.\mathcal{D}\mathbf{b}\mathcal{D}\mathbf{b}'.$$

$$(10.6)$$

With the interaction governed by (10.3), we write

$$S_I[\mathbf{a}(t), \mathbf{b}(t)] = -e_b \int_{\mathbf{\Delta}_+} \mathbf{A}_{\mathrm{ret}}^{(a)}(\mathbf{b}).\dot{\mathbf{b}}dt \qquad (10.7)$$

in (10.6), with a similar expression for $S_I[\mathbf{a}'(t), \mathbf{b}'(t)]$.

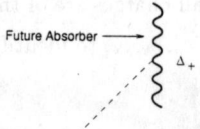

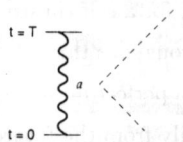

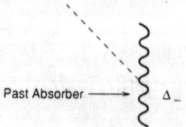

Fig. X.1 The spacetime diagram schematically showing the interaction of an atomic electron with the future absorber.

Because of the cooling produced by the expansion of the universe, b is usually in its ground state. We shall take ψ_i to be the ground state. We can expand the part of the exponential in (10.6) containing the interaction terms only. The unity term in the expansion corresponds to $\psi_i \to \psi_i$. Thus we have

$$
F^{(b)}[\mathbf{a}, \mathbf{a}'] = 1 + \sum_{f \neq i} \int \int \int \int \int \int \psi_f^*(\mathbf{b}_f)\psi_f(\mathbf{b}_f')\psi_i^*(\mathbf{b}_i')\psi_i(\mathbf{b}_i)
$$

$$
\cdot (1/\hbar^2)\left[e_b^2 \int_{\Delta_+} \mathbf{A}_{\mathrm{ret}}{}^{(a)}(\mathbf{b}).\dot{\mathbf{b}}dt \int_{\Delta_+'} \mathbf{A}_{\mathrm{ret}}{}^{(a')}(\mathbf{b}').\dot{\mathbf{b}}'dt \right]
$$

$$
\cdot \exp\left((i/\hbar)\{S_E[\mathbf{b}(t)] - S_E[\mathbf{b}'(t)]\} \right)\mathcal{D}\mathbf{b}'\mathcal{D}\mathbf{b}d^3\mathbf{b}_f d^3\mathbf{b}_i d^3\mathbf{b}_i' d^3\mathbf{b}_f'
$$

$$
+ \text{ terms in } \left[\int_{\Delta_+}\right]^2, \left[\int_{\Delta_+'}\right]^2 + \ldots,
$$

(10.8)

where $\mathbf{A}_{\mathrm{ret}}{}^{(a)}$ is calculated for the path $\mathbf{a}(t)$, $\mathbf{A}_{\mathrm{ret}}{}^{(a')}$ for the conjugate path $\mathbf{a}'(t)$.

Let E_f and E_i be the energies of states ψ_f and ψ_i, respectively. Then first-order perturbation theory shows that the contribution to the above expression from the transition $\psi_i \to \psi_f$ is

$$
\frac{e_b^2}{\hbar^2}\left(\frac{E_f - E_i}{\hbar}\right)^2 M[\mathbf{a}(t)M^*[\mathbf{a}'(t)],
$$

(10.9)

where

$$
M[\mathbf{a}(t)] = \int_{\Delta_+} \exp\left[\frac{i(E_f - E_i)t}{\hbar}\right] dt \int \psi_f^*(\mathbf{b})\mathbf{A}_{\mathrm{ret}}{}^{(a)}(\mathbf{b}).\mathbf{b}\psi_i(\mathbf{b})d^3\mathbf{b}.
$$

(10.10)

To calculate (10.10), we proceed as follows. Take an origin near a to measure the displacement $\mathbf{a}(t)$. Similarly, let $\mathbf{b}(t)$ measure the displacement from an origin near b. Let the origin near b have a relative displacement $\mathbf{R}$ with respect to the origin near a. Then the vector from $\mathbf{a}(t)$ to $\mathbf{b}(t)$ is given by

$$
\mathbf{r} = \mathbf{R} + \mathbf{b} - \mathbf{a}.
$$

(10.11)

For $|\mathbf{R}| \gg |\mathbf{b}|, |\mathbf{a}|$, we get

$$r = |\mathbf{r}| \cong R + \frac{1}{R}\mathbf{R}.(\mathbf{b} - \mathbf{a}). \tag{10.12}$$

In the absence of any dispersion, we would have, in the Coulomb gauge,

$$\mathbf{A}_{ret}^{(a)}(\mathbf{b}) = \frac{e_a}{r - \mathbf{r}.\dot{\mathbf{a}}} \sum_{j=1,2} [\alpha^{(j)}.\dot{\mathbf{a}}]\alpha^{(j)}, \tag{10.13}$$

where $\alpha^{(j)}$ are two unit vectors which form a mutually orthogonal triad with $\mathbf{r}/|\mathbf{r}|$.

The quantity $r - \mathbf{r}.\dot{\mathbf{a}}$ varies only slightly with $\mathbf{b}$, and it is sufficiently accurate to replace it in (10.13) by $R - \mathbf{R}.\dot{\mathbf{a}}$,

$$\mathbf{A}_{ret}^{(a)}(\mathbf{b}) = \frac{e_a}{R - \mathbf{R}.\dot{\mathbf{a}}} \sum_{j=1,2} [\alpha^{(j)}.\dot{\mathbf{a}}]\alpha^{(j)}. \tag{10.14}$$

We Fourier-analyze (10.14) to get

$$\mathbf{A}_{ret}^{(a)}(\mathbf{b}) = \sum_{l=-\infty}^{\infty} \sum_{j=1,2} [\alpha^{(j)}.\mathbf{A}_l]\alpha^j \exp\left(-\frac{2\pi i l t}{T'}\right), \tag{10.15}$$

where

$$\mathbf{A}_l = \frac{e_a}{RT'} \int \frac{\dot{\mathbf{a}}}{1 - \dot{\mathbf{a}}.\mathbf{R}/R} \exp\left(\frac{2\pi i l t'}{T}\right) dt'. \tag{10.16}$$

The range T' of t' is given by

$$t' = t + R + \frac{1}{R}\mathbf{R}.(\mathbf{b} - \mathbf{a}), \qquad 0 \le t \le T. \tag{10.17}$$

Since $\mathbf{b}, \mathbf{R}$, are not to be regarded as varying with t, we have

$$\frac{dt'}{dt} = 1 - \frac{\dot{\mathbf{a}}.\mathbf{R}}{R}, \tag{10.18}$$

so that changing the variable from t' to t in (10.16) gives

$$\mathbf{A}_l = \frac{e_a}{RT'} \exp\left[\frac{2\pi i l}{T'}\left(R + \frac{\mathbf{R}.\mathbf{b}}{R}\right)\right] \int_0^T \dot{\mathbf{a}} \exp\left[\frac{2\pi i l}{T'}\left(t - \frac{\mathbf{a}.\mathbf{R}}{R}\right)\right] dt, \tag{10.19}$$

with

$$T' = T - \frac{\mathbf{R}}{R}[\mathbf{a}(T) - \mathbf{a}(0)]. \tag{10.20}$$

The effect of dispersion in the cosmological medium is to introduce both a phase change and damping into (10.19), modifying it to

$$\mathbf{A}_l = \frac{e_a}{RT'} \exp\left\{-\frac{1}{2}\tau_l + i\left[\frac{2\pi l}{T'}\left(R + \frac{\mathbf{R}.\mathbf{b}}{R}\right) + \chi_l\right]\right\} \int_0^T \dot{\mathbf{a}} \exp\left[\frac{2\pi i l}{T'}\left(t - \frac{\mathbf{a}.\mathbf{R}}{R}\right)\right] dt, \tag{10.21}$$

the phase change being expressed by χ_l and the damping by τ_l.

We now substitute (10.15) into (10.10). We shall make use of the fact (to be shown later) that χ_l is a very large phase angle, and use this to wipe out any cross products of $\mathbf{A}_l, \mathbf{A}_{l'}$, which arise in (10.9), except those with $l' = -l$, giving

$$(e_a^2 e_b^2 / 3R^2 T'^2)((E_f - E_i)^2 / \hbar^4) \sum_{l=-\infty}^{\infty} \left|b_{if}(2\pi l / T')\right|^2 e^{-\tau_l} \left|\int_0^{T'} \exp\left[\frac{it'}{\hbar}\left(E_f - E_i - \frac{2\pi l \hbar}{T'}\right)\right] dt'\right|^2$$

$$\times \sum_{j=1,2} \int_0^T \alpha^{(j)}.\dot{\mathbf{a}} \exp\left[(2\pi l i / T')\left(t - \frac{\mathbf{a}.\mathbf{R}}{R}\right)\right] dt$$

$$\times \int_0^T \alpha^{(j)}.\dot{\mathbf{a}}' \exp\left[2\pi i l / T'\left(\frac{\mathbf{a}'.\mathbf{R}}{R} - t\right)\right] dt, \tag{10.22}$$

in which we have already averaged with respect to all orientations of b to remove product terms in the components of $\alpha^{(1)}$ and $\alpha^{(2)}$, and where $b_{if}\left(\dfrac{2\pi l}{T'}\right)$ is the matrix element of $\mathbf{b} \exp\left[i\dfrac{2\pi l}{T'}\dfrac{\mathbf{R}.\mathbf{b}}{R}\right]$ with respect to ψ_i and ψ_f. We have also ignored variations of $\mathbf{b}$ over $\mathbf{\Delta}+$ in comparison with T'. Expression (10.22) can be simplified further by the results

$$\left|\int_0^{T'} \exp\left[\frac{it'}{\hbar}\left(E_f - E_i - \frac{2\pi \hbar l}{T'}\right)\right] dt'\right|^2 \simeq 2\pi T' \hbar \delta\left(E_f - E_i - \frac{2\pi l \hbar}{T'}\right) \tag{10.23}$$

$$\sum_{i=-\infty}^{\infty} \frac{\sin^2 \alpha}{(\pi l - \alpha)^2} \doteq 1. \tag{10.24}$$

Writing

$$\hbar k = E_f - E_i, \mathbf{k} = \frac{k\mathbf{R}}{R}, \tag{10.25}$$

and using (10.23) and (10.24), we reduce (10.22) to the form

$$(e_a^2 e_b^2 k^2 / 3R^2 \hbar^2) e^{-\tau(k)} |\mathbf{b}_{if}(k)^2|^2 \sum_{j=1,2} \int_0^T \alpha^{(j)} . \dot{\mathbf{a}} e^{-i\mathbf{k}.\mathbf{a}+ikt} dt$$

$$\times \int_0^T \alpha^{(j)} . \dot{\mathbf{a}}' e^{i\mathbf{k}.\mathbf{a}'-ikt} dt = X \text{ (say)}.$$
(10.26)

Suppose that at a distance R in a particular solid angle $d\Omega$, there are $n(k)dk$ particles per unit volume with states f and i such that $(E_f - E_i)/\hbar$ lies in the range k to $k + dk$. Writing $\overline{|\mathbf{b}_{if}(k)|^2}$ for the average of $|\mathbf{b}_{if}(k)|^2$ for all systems satisfying this requirement, the contribution to $F[\mathbf{a}, \mathbf{a}']$ from all absorbers between R and $R + dR$ and in $d\Omega$ is

$$[1 + X]^{n(k)dkR^2 dRd\Omega}. \tag{10.27}$$

Since X is very small and the index is large, we can rewrite (10.27) in the form

$$\exp [X.n(k)dkR^2 dRd\Omega]. \tag{10.28}$$

The function $\tau(k)$ in (10.26) is just the optical depth of the absorbing medium at frequency k. It is not hard to show that

$$\frac{d\tau(k)}{dR} = \frac{4\pi^2}{3} \frac{e_b^2 k}{\hbar} \overline{|\mathbf{b}_{if}(k)|^2} n(k), \tag{10.29}$$

the damping expressed by $\tau(k)$ being due to induced upward transitions in the absorber. Thus (10.29) follows from first-order perturbation theory. Remembering that absorber particles contribute as a product, as in (10.4), we next integrate the exponent of (10.28) with respect to R and with respect to k. Using (10.29), and letting $\tau \to \infty$ as $R \to \infty$, we obtain

$$\exp\left[(e_a^2/4\pi^2\hbar)d\Omega \int_0^\infty kdk \sum_{j=1,2} \int_0^T (\alpha^{(j)}.\dot{\mathbf{a}})\ e^{-i\mathbf{k}.\mathbf{a}+ikt}dt \right.$$

$$\left. \cdot \int_0^T (\alpha^{(j)}.\dot{\mathbf{a}}')e^{i\mathbf{k}.\mathbf{a}'-ikt}dt\right]. \tag{10.30}$$

Integrating finally with respect to Ω, we obtain

$$F[\mathbf{a}(t),\mathbf{a}'(t)] = \exp\left[(e^2/4\pi^2\hbar)\int d\Omega \int_0^\infty kdk \sum_{j=1,2} \int_0^T (\alpha_{\mathbf{k}}^{(j)}\right.$$

$$\left. \cdot \int_0^T (\alpha_{\mathbf{k}}^{(j)}.\dot{\mathbf{a}}')e^{i\mathbf{k}.\mathbf{a}'-ikt}dt\right]. \tag{10.31}$$

The subscript $\mathbf{k}$ has been added to $\alpha^{(j)}$ since we are now integrating with respect to Ω, and the vectors $\alpha^{(j)}$ change as $d\Omega$ changes – $\mathbf{k}$ is a vector in $d\Omega$.

This is the influence functional generated by the future absorber and is the quantum analogue of the radiative damping given by formula (3.7) in Lecture III. Notice that in both cases the final answer depends on quantities related to the local charge a; all parameters relating to the universe have dropped out. However, the important property of 'perfect future absorber' has been used in their derivation. In other words, the final answer desceptively suggests self-action of charge a, when the real effect is of the response of the universe.

C. The Rate of Spontaneous Transition

The last part of the calculation is similar to the procedure that led to (9.7). Expand the exponential in (9.7) and retain only the first-order term in $e^2/\hbar$. We will now drop the suffix a on the charge e_a. Previously we had

$$\frac{e^2}{\hbar^2}\int_0^T \dot{\mathbf{a}}.\mathbf{A}dt \int_0^T \dot{\mathbf{a}}'.\mathbf{A}dt', \tag{10.32}$$

in which $\mathbf{A}$ was a specified field. Now we have the exponent of (10.31). Noting that if $\mathbf{A}$ in (10.32) had not been a real field

$$\frac{e^2}{\hbar^2}\int_0^T \dot{\mathbf{a}}.\mathbf{A}dt \int_0^T \dot{\mathbf{a}}'.\mathbf{A}^*dt' \tag{10.33}$$

would still have led to (9.8), we see that provided we replace $e^2/\hbar^2$ by

$$\frac{e^2}{4\pi^2\hbar} \int d\Omega \int_0^\infty k\,dk \sum_{j=1,2},$$

and provided we write $\mathbf{A} = \alpha_\mathbf{k}^{(j)} e^{-i\mathbf{k}\cdot\mathbf{a}+ikt}$, the present case is the same as the previous one. We therefore obtain

$$P(m \to n) = \frac{e^2}{4\pi^2\hbar}\left(\frac{E_m - E_n}{\hbar}\right)^2 \int d\Omega \int_0^\infty k\,dk \sum_{j=1,2} \left| \int \phi_n^*(\mathbf{a})\mathbf{a}.\alpha_\mathbf{k}^{(j)} e^{i\mathbf{k}\cdot\mathbf{a}} \phi_m(\mathbf{a})d^3\mathbf{a} \right|^2$$

$$\cdot \left| \int_0^T \exp\left[\frac{i}{\hbar}(E_n - E_m + \hbar k)t\right]dt \right|^2.$$

$$(10.34)$$

Using (9.10), we see that $E_m > E_n$ is necessary to obtain a nonzero result, and that the spontaneous emission probability per unit time is

$$(e^2/2\pi\hbar)\ [(E_m - E_n)/\hbar]^3 \int d\Omega \int_0^\infty dk$$

$$(10.35)$$

$$\cdot \sum_{k=1,2} \left| \int \phi_n^*(\mathbf{a})\mathbf{a}.\alpha_\mathbf{k}^{(j)} e^{-i\mathbf{k}\cdot\mathbf{a}} \phi_m(\mathbf{a})d^3\mathbf{a} \right|^2 \delta\left(k - \frac{E_m - E_n}{\hbar}\right)$$

in agreement with the usual expression in standard field theory textbooks.

Averaging (10.35) with respect to solid angle and using the same notation as in (9.21) gives

$$\frac{e^2}{3\pi\hbar}\left(\frac{E_m - E_n}{\hbar}\right)^3 \int_0^\infty dk |\mathbf{a}_{mn}(k)|^2 \delta\left(k - \frac{E_m - E_n}{\hbar}\right) \int d\Omega. \quad (10.36)$$

Now equate the contribution from $d\Omega$ to (9.21) to $\bar{q}(\mathbf{k})$ times the contribution from $d\Omega$ to (10.36). This gives the following definition of $\bar{q}(\mathbf{k})$

$$\bar{q}(\mathbf{k}) = \frac{1}{\hbar}\left(\frac{2\pi}{k}\right)^3 \cdot \frac{1}{2} \sum_{j=1,2} I^{(j)}(\mathbf{k}), \quad (10.37)$$

where $\hbar k = E_m - E_n$ and $I(\mathbf{k})$ is separated into the two polarizations defined in (9.23). In the usual quantum electrodynamics (10.37) is the relation between the field intensity and the average number of quanta per vacuum oscillator in the frequency range k to $k + dk$. Although quanta do not appear explicitly in the present theory, it is interesting that we obtain the usual formulae by taking the spontaneous transition rate as a reference standard. It follows that, if $I(\mathbf{k})$ were to have the value appropriate to a thermodynamic radiation field at temperature T, $\bar{q}(\mathbf{k})$ would follow Planck's law,

$$\bar{q}(\mathbf{k}) \equiv \bar{q}(k) = \frac{1}{\exp\left(\hbar k / T\right) - 1} \tag{10.38}$$

in which the temperature scale has been chosen so that the Boltzmann constant is unity.

The delta-function in (10.35) gives an asymmetry between emission and absorption. Spontaneous transitions are downward because we have taken the absorber particles as being in their ground levels, $E_i \leq E_f$ for all f. We see therefore that the asymmetry of spontaneous emission arises from the assumption of a cold universe. We shall return to this point later.

Exercises

1. Show that spontaneous transitions are inevitable in a universe that has a perfect future absorber and imperfect past absorber.

2. Prove the result of Eq. (10.24).

3. Comment on the fact that the details of opacity of the universe drop out in the final answer for the spontaneous transition probability.

4. Compare the field quantization method for deriving the spontaneous transition probability with the method used here.

LECTURE XI : THE COMPLETE INFLUENCE FUNCTIONAL AND THE LEVEL SHIFT FORMULA

A. The Influence Functional

The previouse lecture derived the specific result of the downward transition of an atomic electron without any local direct particle field. The downward transition may be seen as the probabilistic outcome of the various paths available to the electron, each path invoking a response from the future absorber. This cycle of events

Motion of the electron → Disturbance in the absorber

↑   ↓

Disturbance of the electron ← Response of the absorber

has to be self-consistent and requires that the electron jumps 'down' rather than 'up' spontaneously. The role of the absorber is to *receive* energy from the electron and hence to cause its *downward* transition. Had the past absorber of the universe instead, been efficient the situation would have turned out the other way leading to spontaneous *upward* transitions.

In conventional quantum field theory this result is derived by *postulating* commutation relations for the electromagnetic field operators and then defining the creation and annihilation operators acting on states with n photons. A creation operator changes the state to one of $n + 1$ photons while the annihilation operator changes it to one of $n - 1$ photons. The assymetry introduced between the creation and annihilation operators ultimately gets translated into the one sided (downward) transition of the electron. Clearly, the relationship to cosmological asymmetry in the direct particle theory is physically more understandable than the formal field theoretic approach.

However, it is necessary to generalize the treatment of the previous lecture to cover a wider range of physical situations. To this end we will begin by taking a fresh look at the influence functional.

The complete influence functional was given by (10.8), of which the two terms in the last line remain to be discussed.

$$-(1/2\hbar^2)\int\int\int S_I^2[\mathbf{a}(t),\mathbf{b}(t)]\exp\{iS_0[\mathbf{b}(t)]/\hbar\}\psi_i^*(\mathbf{b}_f)\psi_i(\mathbf{b}_i)d^3\mathbf{b}_f d^3\mathbf{b}_i \mathcal{D}^3\mathbf{b}(t),$$

$$-(1/2\hbar^2)\int\int\int S_I^2[\mathbf{a}'(t),\mathbf{b}'(t)]\exp\{-iS_0[\mathbf{b}'(t)]/\hbar\}\psi_i(\mathbf{b}'_f)\psi_i^*(\mathbf{b}'_i)d^3\mathbf{b}'_f d^3\mathbf{b}'_i \mathcal{D}^3\mathbf{b}'(t).$$

$$(11.1)$$

It will be sufficient to work out the first of these terms, since the second can then be written down by inspection.

Inserting (10.15) for $\mathbf{A}_{\text{ret}}^{(a)}$ in the expression (10.7) for S_I leads to

$$-(e^2/2\hbar^2)\sum_{l=-\infty}^{\infty}\int\int\int\int_0^{T'}(\mathbf{A}_l.\dot{\mathbf{b}})e^{-2\pi i l t'/T}dt'\int_0^{T'}(\mathbf{A}_l^*.\dot{\mathbf{b}})e^{2\pi i l t'/T}dt'$$

$$(11.2)$$

$$.\exp\{iS_0[\mathbf{b}(t)]/\hbar\}\psi_i^*(\mathbf{b}_f)\psi_i(\mathbf{b}_i)d^3\mathbf{b}_f d^3\mathbf{b}_i \mathcal{D}^3\mathbf{b}(t).$$

Using ordinary perturbation methods for second-order transitions, (11.2) can be reduced to

$$-(e^2/\hbar^2)\sum_{l=-\infty}^{\infty}\sum_g[(E_i-E_g)/\hbar]^2\left|\int\psi_i^*(\mathbf{b})(\mathbf{b}.\mathbf{A}_l)\psi_g(\mathbf{b})d^3\mathbf{b}\right|^2$$

$$.\int_0^{T'}\exp\left[\frac{i}{\hbar}\left(E_i-E_g-(2\pi l\hbar/T')\right)t'\right]dt'$$

$$(11.3)$$

$$.\int_0^{t'}\exp\left[(i/\hbar)\left(E_0-E_i+(2\pi l\hbar/T')\right)\tilde{t}'\right]d\tilde{t}',$$

where the summation with respect to g is over all intermediate states of b.

Now insert (10.21) for $\mathbf{A}_l$ in (11.3), to give

$$(e^2/3\hbar^2 R^2 T'^2) \sum_g \left(\frac{E_g - E_i}{\hbar}\right)^2 \sum_{l=-\infty}^{\infty} |\mathbf{b}_{ig}(2\pi l/T')|^2 e^{-\tau_l}$$

$$\cdot \int_0^{T'} \exp\left[(i/\hbar)\left(E_i - E_g - \frac{2\pi\hbar l}{T'}\right)t'\right]dt'$$

$$\cdot \int_0^{t'} \exp\left[(i/\hbar)\left(E_g - E_i + 2\pi\hbar l/T'\right)\tilde{t}'\right]d\tilde{t}' \tag{11.4}$$

$$\cdot \sum_{j=1,2} \int_0^T \alpha^{(j)} . \dot{\mathbf{a}} \exp\left[(2\pi i l/T')\left(t - \frac{\mathbf{a}.\mathbf{R}}{R}\right)\right]dt$$

$$\cdot \int_0^T \alpha^{(j)} . \dot{\mathbf{a}} \exp\left[-(2\pi i l/T')\left(\tilde{t} - \frac{\mathbf{a}.\mathbf{R}}{R}\right)\right]d\tilde{t}$$

after averaging with respect to the orientation of b.

Although (11.4) appears complicated we shall find simplifications. For fixed ψ_g, the main contributions come from $2\pi/\hbar/T' \cong E_i - E_g$. Only when this condition is satisfied can the integrals in the second and third lines of (11.4) yield a contribution that behaves as T'^2. Defining

$$\mathbf{k} = \frac{E_i - E_g}{\hbar} . \frac{\mathbf{R}}{R}, \tag{11.5}$$

we can therefore write

$$-\frac{e^2}{3\hbar^2 R^2 T'^2} \sum_g k^2 |\mathbf{b}_{ig}(k)|^2 . e^{-\tau(k)} \int_0^{T'} e^{-ikt'} dt' \int_0^{t'} e^{ik\tilde{t}'} d\tilde{t}'$$

$$\cdot \sum_{j=1,2} \int_0^T \alpha^{(j)} . \dot{\mathbf{a}} e^{-i\mathbf{k}.\mathbf{a}} dt \int_0^T \alpha^{(j)} . \dot{\mathbf{a}} e^{i\mathbf{k}.\mathbf{a}} d\tilde{t} \sum_{l=-\infty}^{\infty} \exp\left[\frac{2\pi i l}{T'}(t - \tilde{t} - t' + \tilde{t}')\right]. \tag{11.6}$$

Furthermore

$$\sum_{l=-\infty}^{\infty} \exp\left[\frac{2\pi i l}{T'}(t - \tilde{t} - t' + \tilde{t}')\right] = T'\delta(t - \tilde{t} - t' + \tilde{t}'). \tag{11.7}$$

We must therefore have $t \geq \tilde{t}$ since $t' \geq \tilde{t}'$, and (11.6) becomes

$$-\frac{e^2}{3\hbar^2 R^2} \sum_g k^2 e^{-\tau(k)} |\mathbf{b}_{ig}(k)|^2 \sum_{j=1,2} \int_0^T \alpha^{(j)} e^{i\mathbf{k}.\mathbf{a} + ik\tilde{t}} d\tilde{t}. \tag{11.8}$$

This contribution to $F^{(b)}$ must be added to (10.31).

As before it is sufficient to consider a single state g, since all states were automatically included in the discussion that followed (10.31). Indeed the summation with respect to all absorbers proceeds exactly as before, and in place of (10.31) we now have

$$
\begin{aligned}
F[\mathbf{a}(t), \mathbf{a}'(t)] = \ &\exp\left[\frac{e^2}{4\pi^2\hbar}\int d\Omega \int_0^\infty kdk\right. \\
&\times \sum_{j=1,2}\left\{\int_0^T (\alpha_{\mathbf{k}}^{(j)}.\dot{\mathbf{a}})e^{-i\mathbf{k}.\mathbf{a}+ikt}dt\int_0^T(\alpha_{\mathbf{k}}^{(j)}.\dot{\mathbf{a}}')e^{+i\mathbf{k}.\dot{\mathbf{a}}'-ikt'}dt'\right. \\
&-\int_0^T(\alpha_{\mathbf{k}}^{(j)}.\dot{\mathbf{a}})e^{-i\mathbf{k}.\mathbf{a}-ikt}dt\int_0^t(\alpha_{\mathbf{k}}^{(j)}.\dot{\mathbf{a}})e^{i\mathbf{k}.\dot{\mathbf{a}}+ik\tilde{t}}d\tilde{t} \\
&\left.\left.-\int_0^T(\alpha_{\mathbf{k}}^{(j)}.\dot{\mathbf{a}}')e^{-i\mathbf{k}.\mathbf{a}'+ikt}dt\int_0^t(\alpha_{\mathbf{k}}^{(j)}.\dot{\mathbf{a}}')e^{i\mathbf{k}.\dot{\mathbf{a}}'-ik\tilde{t}}d\tilde{t}\right\}\right],
\end{aligned}
\tag{11.9}
$$

in which we have also included the second term of (11.1).

The expression (11.9) is an influence functional and it obeys the general rules discussed by Feynman and Hibbs (9), p. 347). If we identify the paths, $\mathbf{a}(t) \equiv \mathbf{a}'(t)$, the exponent in (11.9) vanishes. The paths do not act on themselves via the response of the Universe.

B. The Level Shift Formula

We now discuss another consequence of this complete influence functional of (11.9). For, we will discover that the two extra terms we have now included also produce an observable effect in that they lead to slight changes of the electron energy levels.

The new terms in (11.9) have no effect on the calculation of $P(m \to n), m \neq n$, but they are necessary to obtain $P(m \to m)$. We now show that

$$
P(m \to m) = 1 - \sum_{n \neq m} P(m \to n).
\tag{11.10}
$$

$P(m \to m)$ is given by writing m for n in (9.1). Expanding the exponential in (11.9) to first order we have

$$P(m \to m) = 1 - (e^2/4\pi^2\hbar) \int \int \int \phi_m^*(\mathbf{a}_f) \exp\left\{\frac{i}{\hbar}S_0[\mathbf{a}(t)]\right\}\phi_m(\mathbf{a}_i) \int d\Omega \int_0^\infty k\,dk$$

$$\cdot \sum_{j=1,2} \int_0^T (\dot{\mathbf{a}}.\alpha_\mathbf{k}^{(j)})e^{-i\mathbf{k}.\mathbf{a}-ikt}dt \int_0^t (\dot{\mathbf{a}}.\alpha_\mathbf{k}^{(j)})e^{i\mathbf{k}.\mathbf{a}+ik\tilde{t}}d\tilde{t}\mathcal{D}^3\mathbf{a}(t)d^3\mathbf{a}_f d^3\mathbf{a}_i$$

$$-(c^2/4\pi^2\hbar)\int \int \int \phi_m(\mathbf{a}'_f) \exp\left\{-\frac{i}{\hbar}S_0[\mathbf{a}'(t)]\right\}\phi_m^*(\mathbf{a}'_i) \int d\Omega \int_0^\infty k\,dk$$

$$\cdot \sum_{j=1,2} \int_0^T (\dot{\mathbf{a}}'.\alpha_\mathbf{k}^{(j)})e^{-i\mathbf{k}.\mathbf{a}'+ikt}dt$$

$$\cdot \int_0^t (\dot{\mathbf{a}}'.\alpha_\mathbf{k}^{(j)})e^{i\mathbf{k}.\mathbf{a}'-ik\tilde{t}}d\tilde{t}\mathcal{D}^3\mathbf{a}'(t)d^3\mathbf{a}'_f d^3\mathbf{a}'_i.$$

$$(11.11)$$

The term involving $\mathbf{a}(t)$ separates from the term involving $\mathbf{a}'(t)$. The path integrals are not hard to evaluate. We obtain

$$P(m \to m) = 1 - (e^2/4\pi^2\hbar) \sum_{j=1,2} \int d\Omega \int_0^\infty k\,dk$$

$$\cdot \sum_n \left((E_n - E_m)/\hbar\right)^2 \left|\int \phi_m^*(\mathbf{a})\alpha_\mathbf{k}^{(j)}.\mathbf{a}e^{-i\mathbf{k}.\mathbf{a}}\phi_n(\mathbf{a})d^3\mathbf{a}\right|^2$$

$$(11.12)$$

$$\cdot \int_0^T dt \int_0^t \left[\exp\left\{\frac{i}{\hbar}(E_m - E_n - \hbar k)(t - \tilde{t})\right\}\right.$$

$$\left.+\exp\left\{\frac{i}{\hbar}(E_m - E_n - \hbar k)(\tilde{t} - t)\right\}\right]d\tilde{t}.$$

The term in $\exp\{i/\hbar(E_m - E_n - \hbar k)(t - \tilde{t})\}$ in (11.12) comes from the quadratic term in the path $\mathbf{a}(t)$ in $F[\mathbf{a}(t), \mathbf{a}'(t)]$, and the term in $\exp[i/\hbar(E_m - E_n - \hbar k)(\tilde{t} - t)]$ comes from the quadratic term in $\mathbf{a}'(t)$.

The last two integrals of (11.12) give

$$2\int_0^T \frac{\sin[E_m - E_n)/\hbar - k]t}{(E_m - E_n)/\hbar - k}dt \cong 2\pi T\delta\left(k - \frac{E_m - E_n}{\hbar}\right). \qquad (11.13)$$

Hence (11.12) is just $1 - \sum_{E_n < E_m} P(m \to n)$. Since $P(m \to n)$ is zero for $E_n > E_m$ we obtain (11.10). Probability is therefore conserved.

The system a is by hypothesis in the state m at $t = 0$. The effect of the response of the universe is to change the amplitude for the system to be in the state m at time T from $\phi_m \exp(-iE_m T/\hbar)$ to

$$\phi_m \exp\left[-\frac{i}{\hbar}(E_m + \Delta E_m) - \frac{1}{2}\gamma\right]T \cong \left(1 - \frac{\gamma T}{2} - \frac{i\Delta E_m T}{\hbar}\right)\phi_m e^{iE_m T/\hbar} \qquad (11.14)$$

for T not too large. The probability of the system being in the state m at time T is therefore

$$P(m \to m) \cong \left(1 - \frac{\gamma T}{2} - \frac{i\Delta E_m T}{\hbar}\right)\left(1 - \frac{\gamma T}{2} + \frac{i\Delta E_m T}{\hbar}\right)$$

$$= 1 - \left(\frac{\gamma}{2} + \frac{i\Delta E_m}{\hbar}\right)T - \left(\frac{\gamma}{2} - \frac{i\Delta E_m}{\hbar}\right)T + O(\Delta E_m^2). \qquad (11.15)$$

Suppose we identify the second and third terms on the right hand side of (11.15) with the second and third terms on the right hand side of (11.11). Then

$$\left(\frac{1}{2}\gamma + i\,\frac{\Delta E_m}{\hbar}\right)T = \frac{e^2}{4\pi^2\hbar}\int\int\int \phi_m^*(\mathbf{a}_f)\phi_m(\mathbf{a}_i)\exp\left\{\frac{i}{\hbar}S_0[\mathbf{a}(t)]\right\}\int d\Omega \int_0^\infty k\,dk$$

$$\cdot \sum_{j=1,2}\int_0^T \dot{\mathbf{a}}.\alpha_{\mathbf{k}}^{(j)} e^{-i\mathbf{k}.\mathbf{a}-ikt}\,dt \int_0^t \dot{\mathbf{a}}.\alpha_{\mathbf{k}}^{(j)} e^{+i\mathbf{k}.\mathbf{a}+ik\tilde{t}}\,d\tilde{t}\,\mathcal{D}^3\mathbf{a}(t)\,d^3\mathbf{a}_f\,d^3\mathbf{a}_i. \qquad (11.16)$$

Care is needed in evaluating the path integral because a "cross-over" term arises at $t = \tilde{t}$. This term yields

$$\frac{ie^2 T}{\pi m}\int_0^\infty k\,dk, \qquad (11.17)$$

while the main term in the reduction is

$$(e^2/4\pi^2\hbar) \sum_{j=1,2} \int d\Omega \int_0^\infty kdk \sum_n \left((E_n - E_m)/\hbar\right)^2 \left| \int \phi_n^*(\mathbf{a})\mathbf{a}.\alpha_\mathbf{k}^{(j)} e^{-i\mathbf{k}.\mathbf{a}} \phi_m(\mathbf{a}) d^3\mathbf{a} \right|^2$$

$$\cdot \int_0^T dt \int_0^t \exp\left\{\frac{i}{\hbar}(E_m - E_n - \hbar k)(t - \tilde{t})\right\} d\tilde{t}$$

$$= \frac{1}{2} \sum_{E_n < E_m} P(m \to n) + (ie^2/4\pi^2\hbar) \sum_{j=1,2} \int d\Omega \int_0^\infty kdk$$

$$\cdot \sum_n \left((E_n - E_m)/\hbar\right)^2 \left| \int \phi_n^*(\mathbf{a})\mathbf{a}.\alpha_\mathbf{k}^{(j)} e^{-i\mathbf{k}.\mathbf{a}} \phi_m(\mathbf{a}) d^3\mathbf{a} \right|^2$$

$$\cdot \cdot \int_0^T dt \int_0^t \sin\left[(E_m - E_n - \hbar k)(t - \tilde{t})/\hbar\right] d\tilde{t}. \tag{11.18}$$

Collecting terms

$$\frac{\gamma}{2} + \frac{i}{\hbar}\Delta E_m = \frac{1}{2T} \sum_{E_n < E_m} P(m \to n) + \frac{ie^2}{\pi m} \int_0^\infty kdk$$

$$+ \frac{ie^2}{4\pi^2} \sum_n \left(\frac{E_m - E_n}{\hbar}\right)^2 \sum_{j=1,2} P.P. \int_0^\infty \frac{kdk}{E_m - E_n - \hbar k} \int d\Omega \tag{11.19}$$

$$\cdot \left| \int \phi_n^*(\mathbf{a})\alpha_\mathbf{k}^{(j)}.\mathbf{a} e^{-i\mathbf{k}.\mathbf{a}} \phi_m(\mathbf{a}) d^3\mathbf{a} \right|^2.$$

The summation in the last term is not restricted to $E_n < E_m$.

C. The Radiation Cut off at the Absorber

The cosmological features of the response of the universe was discussed in Lecture VI, where Table II gave results for the classical electromagnetic theory. Of the cosmological models discussed in the literature only the steady state and quasi-steady state theories met the requirements of the absorber theory, and the same may be anticipated in the present quantum version of the absorber theory. There may well be other cosmological models still not examined in detail that also meet the same requirements. Indeed any model with a proper density of matter that does not fall below some non-zero lower limit along the future light cone is expected to satisfy the required response condition. Here we consider a further feature of the absorber theory using the steady-state model as the simplest example of such a class of models.

The work of Lectures IX–XI was in flat spacetime. This work can be taken over to cosmological spaces of the Robertson-Walker type by making a conformal transformation of the latter, as in Lecture V. There is, however, the slight notational problem that in the discussion of Lectures IX–XI, the time was denoted by t, whereas in Lecture V the time t was in the cosmological space and τ in the flat conformal space. To retain the explicit formulae of Lectures IX–XI we therefore interchange t and τ as they appeared in Lecture V. For an observer at $t = 0, r = 0$ the line element of the steady-state model then takes the form

$$ds^2 = (1 - Ht)^{-2}[dt^2 - dr^2 - r^2(d\theta^2 + \sin^2\theta d\phi^2)], \qquad (11.20)$$

where $H \simeq 3.10^{-18}s^{-1}$ is a constant of the theory. This model is illustrated in Fig. XI.1.

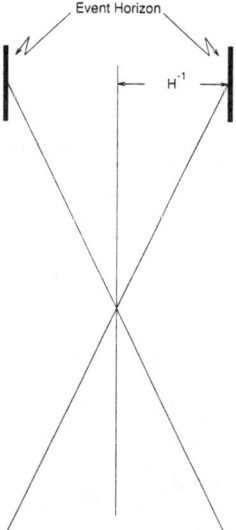

Fig. XI.1 In the conformally flat form of the metric of steady state cosmology the light cones are 'straight'. The event horizon in the future is given at $r = H^{-1}$, shown above by a thick boundary. The frequency of radiation emitted from $r = 0, t = 0$ becomes zero when $r \to H^{-1}$.

In the spontaneous transition $E_m \rightarrow E_n$ of the local system, the important frequencies involved in the influence functional (11.9) are those in the neighbourhood of

$$k = (E_m - E_n)/\hbar. \tag{11.21}$$

This frequency has to be matched by the value of $(E_f - E_i)/\hbar$ for the absorber transition $\psi_i \rightarrow \psi_f$. If, however, we wish to consider E_i, E_f, with respect to proper time at the absorber, it is necessary to take account of the redshift effect of the expansion of the universe. Thus when a wave, starting with frequency k at the source, reaches an absorber at coordinate r, the proper frequency has become

$$\omega = (1 - Hr)k, \tag{11.22}$$

and E_i, E_f, with respect to local proper time are related to k by

$$E_f - E_i = \hbar k(1 - Hr). \tag{11.23}$$

If we displace the absorber by a proper distance dl away from the source, the frequency in tune with the wave changes by $d\omega$, where

$$d\omega = -Hk\,dr, \qquad dr = (1 - Hr)dl. \tag{11.24}$$

The second relation in (11.24) follows from the conformal transformation. Hence we get

$$\frac{d\omega}{\omega} = -H\,dl. \tag{11.25}$$

Suppose the absorbers are effective over a range $[\omega_{\min}, \omega_{\max}]$ of frequencies. Then if $k > \omega_{\max}$, the range of l over which the absorbers can be in tune with k is

$$l = \int dl = H^{-1} \int_{\omega_{\min}}^{\omega_{\max}} \frac{d\omega}{\omega} = H^{-1} \ln \frac{\omega_{\max}}{\omega_{\min}}. \tag{11.26}$$

For complete absorption, we need $l \to \infty$. Hence we need a process with $\omega_{min} \to 0$. Collisional absorption provides such a process, and will be used from hereon, although any process with $\omega_{min} \to 0$ will suffice equally well.

At the low frequencies at which the main collisional absorption occurs it is sufficient to use classical considerations, according to which the alteration of a wave of frequency ω over a length dl caused by collisions with frequency ν is

$$\exp\left[-\frac{2\pi\nu N e^2}{m\omega^2}dl\right], \tag{11.27}$$

where N is the number density of electrons, ϵ the electronic charge, and m the electronic mass. Using (11.22) and (11.24), the damping produced over the coordinate range $0 \le r \le R$ is

$$\exp\left[-\frac{2\pi\nu N e^2}{mk^2}\int_0^R \frac{dr}{(1-Hr)^3}\right]. \tag{11.28}$$

For large R, appreciable damping occurs when

$$(1-Hr)^2 \simeq \frac{2\pi\nu N e^2}{mk^2 H}, \tag{11.29}$$

which corresponds to an effective proper frequency

$$\omega_{eff}^2 \simeq \frac{2\pi\nu N e^2}{mH}. \tag{11.30}$$

For ionized hydrogen, ν at ω_{eff} is given by

$$\nu = 2Nv\left(\frac{e^2}{mv^2}\right)^2 ln\left(\frac{mv^2}{\hbar\omega_{eff}}\right), \tag{11.31}$$

where v is a typical electron veleocity. Taking $H^{-1} \sim 3.10^{17}$ sec, $N \sim 10^{-9}$ cm^{-3}, $v = 1/300$ (of velocity of light), and substituting for $e, m, \hbar$ in (11.30) and (11.31) we can solve for ω_{eff} and ν,

$$\omega_{eff} \simeq 80 \text{ sec}^{-1}, \quad \nu = 1.3 \times 10^{-14} \text{ sec}^{-1}. \tag{11.32}$$

Thus a wave with frequency greater than 10^2 s^{-1} is first redshifted to $\sim 10^2$ s^{-1} and then absorbed. The effective absorption takes place over a proper distance of the order of 10^{28} cm. A wave with frequency less than 10^2 s^{-1} is absorbed without having to be redshifted.

From (11.32) we note that the dimensionless paramter

$$\frac{4\pi Ne^2}{m\omega_{\text{eff}}^2} \simeq 5.10^{-4}, \tag{11.33}$$

while the real part n of the refractive index given by $1 - 2\pi Ne^2/m\omega_{\text{eff}}^2$ is little different from unity, showing that ω_{eff} is appreciably larger than the plasma frequency $(4\pi Ne^2/m)^{1/2}$, showing also that the wave is effectively absorbed well before the redshift causes the frequency to fall so low that the real part of the refractive index becomes negative, with the wave no longer able to propagate.

From this deduction two important inferences can be made. One explains the situation at (10.22) where random phasing was used to average cross-products of A_l and $A_{l'}$, $l \neq -l'$, to zero. Consider two waves with slightly different values of the initial frequency, k and $k + dk$. From (11.29), absorption occurs at r and $r + dr$ with

$$H\,dr = \omega_{\text{eff}}.dk/k^2. \tag{11.34}$$

From the line element (11.20) the proper distance associated with this coordinate displacement is $(1 - Ht)^{-1}dr$, and with $r = t$ along a light ray, this is the same as $(1 - Hr)^{-1}dr$. Using (11.33) and $\omega_{\text{eff}} = (1 - Hr)k$ this proper distance is

$$H^{-1}dk/k \simeq 3.10^{17}dk/k \text{ sec} \simeq 10^{28}dk/k \text{ cm}. \tag{11.35}$$

For $k \simeq 10^{15}s^{-1}$ at optical frequencies this is a large distance even for dk as small as $1s^{-1}$. The proper distance between the particles absorbing the wave of frequency k and that of frequency $k + dk$ is very large compared with the interparticle distance. The responses to k and to $k+dk$ are therefore uncorrelated in phase and the cancelling of waves with $l' = -l$ at (10.22) is explained. The same issue arises in the usual version of quantum mechanics. How is random phasing between the degrees of freedom of the quantised field to be understood? So far as we are aware it is assumed and not understood.

The second inference that will assume importance in later lectures is that a frequency cut-off now emerges cosmologically.

In the above discussion we wrote that emission at $r = 0$ occurs for $t \simeq 0$, not $t = 0$, since some finite interval of t is needed for emission to take place. The duration of emission for spontaneous emission was $0 \leq t \leq T$ with $T \gg k^{-1}$, this condition being necessary to establish the delta-function property at (9.10). The early oscillations emitted immediately after $t = 0$ are absorbed by particles at $t = r$ with

$$r = H^{-1}(1 - \omega_{\mathrm{eff}}/k), \tag{11.36}$$

while the later oscillations emitted at $t = T$ are absorbed at $t = r$ with

$$r = T + H^{-1}(1 - \omega_{\mathrm{eff}}/k). \tag{11.37}$$

Since, however, the cosmological model has an horizon cut-off at $t = H^{-1}$ we cannot have $t = r$ with r given by (11.37) unless

$$k < \omega_{\mathrm{eff}} H^{-1}/T. \tag{11.38}$$

For $\omega_{\mathrm{eff}} = 80$ s^{-1}, $H^{-1} = 3.10^{17}$s, $T = 10^{-12}$s as an example, the cut-off on k is at 2.4×10^{31} s^{-1}. Above this limit the response condition fails. It does so for the same value of k for every observer in the cosmological Hubble flow, provided every observer examines the same radiation process with the same T.

We have thus deduced the existence of a very high frequency cut-off. It has no practical relevance for the calculation of transition probabilities. But at a later stage, when we come to the radiative correction process in relativistic quantum electrodynamics, it will be of critical importance to the logical development of the theory. It avoids the infinities of quantum electrodynamics.

Exercises

1. Discuss how the electron energy levels in an atom shift through its interaction with the universe.

2. Prove the relation (11.7). What is its continuum version?

3. The numbers derived in §C relate to specific observed values of cosmological parameters, H and N. Change these values by a factor 100 either way to see how sensitive T is to these variations.

PART   III

RELATIVISTIC QUANTUM
ELECTRODYNAMICS

LECTURE XII : PATH INTEGRALS FOR RELATIVISTIC PARTICLES

A. Introduction

The work of the preceding section has amply demonstrated that the Wheeler-Feynman absorber theory of radiation can be extended into the quantum domain. The explanation of the phenomenon of spontaneous transition was hitherto considered to demand a quantum field theory. The degrees of freedom vested in the electromagnetic field make the quantum vacuum non-trivial and therefore the atomic electron is supposed to jump down the energy ladder even in the absence of the external field, because it interacts with the field vacuum.

In the action at a distance picture, the role of the vacuum is taken over by the response of the universe. The calculation of the transition probability in the previous discussion has shown that provided we live in the right type of universe (perfect future absorber and imperfect past absorber) the answer comes out right. Thus we could argue that a quantum field theory is sufficient but not necessary for understanding spontaneous transition. Can action at a distance replace quantum field theory altogether just as it can demonstrably replace the classical field theory?

The answer is not immediately obvious. First of all, spontaneous transition represents but the tip of the iceberg of phenomena coming under the purview of quantum field theory. What about phenomena like Compton scattering, pair creation and annihilation, vacuum polarization, self energy effects like the Lamb shift, etc. ? Unless the full gamut of quantum electrodynamic results are described by the action at a distance we cannot look upon it as a viable alternative to field theory.

We will therefore attempt to review the progress made in the above direction. This necessitates, however, the development of a formalism for path integrals for relativistic particles. First we will discuss the motion of a single relativistic fermion, i.e., a Dirac particle in terms of Feynman's path integral formalism suitably extended. We will then discuss a system containing many fermions interacting via the electromagnetic action at a distance a la Fokker. This will bring in the response of the universe and the quantum analogue of the Dirac formula (2.2) for radiative reaction. We will then consider the issues relating to self-action and renormalization.

B. The Motion of a Dirac Particle

We first summarize the path integral approach from Lecture 7 in a form that will be adaptable to the relativistic case.

Consider the motion of an electric charge a with mass m_a and charge e_a moving freely and nonrelativistically from a spacetime point 1 to a spacetime point 2. Let us assume that the spacetime coordinates of 1 and 2 are respectively $(\mathbf{a}_1, t_1)$ and $(\mathbf{a}_2, t_2)$. In Newtonian mechanics this particle would move along a definite path Γ_c in spacetime connecting point 1 to point 2. This is the path of zero acceleration. In quantum mechanics, as discussed in the last section, there is no such unique path but a whole range of paths Γ all starting at 1 and ending at 2. The overall motion of the particle from 1 to 2 is described by a propagator $K[2; 1]$ that is obtained by summing the probability amplitudes along all the paths according to the formula:

$$K[2; 1] = \int \exp \{iJ[\Gamma]\}\mathcal{D}\Gamma \tag{12.1}$$

where $J[\Gamma]$ is the classical action computed for path Γ. The path integral can be evaluated and the answer is

$$K[2; 1] = \left[\frac{m_a}{2\pi i(t_2 - t_1)}\right]^{3/2} \exp \left\{\frac{im_a(\mathbf{a}_2 - \mathbf{a}_1)^2}{2(t_2 - t_1)}\right\}.\theta(t_2 - t_1). \tag{12.2}$$

Here θ is the Heaviside function. [The reader is reminded that we have taken $\hbar = 1, c = 1$.] The propagator $K[2; 1]$ satisfies the well known Schrödinger equation

$$\left[\frac{\partial}{\partial t_2} + \frac{1}{2im_a}\nabla_2^2\right] K[2; 1] = \delta_4(2; 1). \tag{12.3}$$

Instead of using (12.1) we could proceed in the following way. Along the typical path Γ mark points $X_i, i = 0, 1, 2, \ldots N$, with the end points 1 and 2 corresponding to X_0 and X_N respectively. With N sufficiently large, we can consider a typical segment $X_i X_{i+1}$ as infinitesimal. Then define $P(\Gamma)$ as the product

$$P(\Gamma) = \prod_{i=1}^{N} A_i^{-1} K[X_i; X_{i-1}], \tag{12.4}$$

where the A_i are normalizing constants. Proceeding to a limit as $N \to \infty$, we can recover the expression Eq.(12.1). In a sense Eq.(12.1) is the inverse of Eq.(12.4). For details we refer the reader to Feynman's original paper (1949) or the classic book by Feynman and Hibbs (1965).

Another useful result relates the particle propagator to the complete set $\{u_n\}$ of normalized stationary eigen-solutions of the homogeneous Schrödinger equation:

$$K[2;1] = \sum_n u_n(2)\bar{u}_n(1)\theta(t_2 - t_1). \tag{12.5}$$

Note that the propagator is time-asymmetric in the sense that it assigns a zero probability amplitude for motion backwards in time while the full set of eigen-solutions is used to describe the amplitude for forward propagation. In the discussion of the relativistic motion to be considered next, this aspect undergoes a serious modification.

The classical action for a relativistic particle of rest mass m_a is given by

$$J = - \int m_a da. \tag{12.6}$$

This action, however, does not describe a fermion like an electron because it contains no information on the intrinsic spin. Rather than look for a classical action containing this information we will follow the alternative procedure of Eq.(12.4). For, we already know that the wave equation generalizing Eq.(12.3) is the Dirac equation which for the propagator $K[2;1]$ becomes

$$(\nabla\!\!\!\!/_2 + im_a)K_0[2;1] = \delta_4(2,1). \tag{12.7}$$

For reasons to be made clear shortly, we have distinguished the propagator by a suffix 0. In analogy with the nonrelativistic limit, $K_0[2;1]$ is expected to satisfy the temporal condition

$$K_0[2;1] = 0 \quad , \quad \text{for } t_2 < t_1. \tag{12.8}$$

However, here we run into the well known problem of negative energy states. If we express the solution of Eq.(12.7) in terms of an expression like (12.5) we find that the complete set $\{u_n\}$ has to include negative energy solutions also. Dirac had sought to get round the problem via the "hole theory", thus effectively converting it into a many-particle problem.

Feynman (1949) got round the problem of negative energy states by redefining the propagator solution of Eq.(12.7) in the following way:

$$K_+[2;1] = \begin{cases} \displaystyle\sum_{E_n>0} u_n(2)\bar{u}_n(1), & t_2 > t_1, \\ \displaystyle\sum_{E_n<0} u_n(2)\bar{u}_n(1), & t_2 < t_1. \end{cases} \tag{12.9}$$

Notice that the propagator $K_+[2;1]$ allows propagation *backwards* as well as forward in time. The positive energy states ($E_n > 0$) contribute to forward propagation while the negative energy states ($E_n < 0$) contribute to backward propagation. In standard language we say that a negative energy electron going backward in time corresponds to a positive energy positron going forward in time. We shall refer to $K_+[2;1]$ as the Feynman propagator.

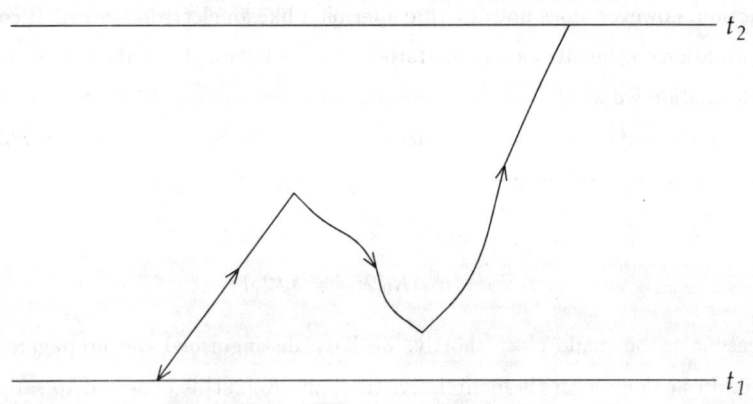

Fig. XII.1 Paths turning backwards in time like that shown here can have non-zero amplitude.

Let us now see how all this affects our definition of probability amplitude if we seek to generalize Eq.(12.4) with the Feynman propagator K_+ replacing K_0. One difference is immediately apparant. In the nonrelativistic case a path that turned backwards in time would automatically have zero probability amplitude. Not so any more. We could now have paths as shown in Fig. XII.1 with nonzero amplitudes.

How do we interpret these paths? In principle they tell us that a forward (in time) moving electron can be scattered to go backwards in time and vice versa. This with Feynman's reinterpretation corresponds to the phenomena of pair annihilation or creation which can happen with or without any external electromagnetic disturbance. We will next consider how these cases are to be looked at in the action at a distance picture.

C. Motion in an External Potential

Given the Feynman propagator for a free particle a, we next ask for a quantum mechanical description of a charge e_a moving under the external electromagnetic potential A_i of other electric charges. Instead of the propagator of Eq.(12.9) we now have another denoted by $K_+^A[2; 1]$ which satisfies the equation

$$(\nabla_2 + ie_a \, A\!\!\!/\,(2) + im_a) K_+^A[?; 1] - \delta_4(2,1). \tag{12.10}$$

Both in Eq.(12.7) and (12.10) a suitable limiting process is used to define the derivatives of the K_0 or K_+ propagators.

The result proved by Feynman, using second quantization of the particle wavefunction [cf. Feynman 1949, Appendix] was that we can ignore the hole theory in an amplitude calculation provided we use $K_+^A[2; 1]$ and multiply the amplitude by C_v, C_v being the amplitude for vacuum to remain a vacuum. C_v is given by

$$C_v = \exp(-L), \quad L - \sum_{n \geq 2} L^{(n)}, \tag{12.11}$$

where $L^{(n)}$ is the amplitude for the occurrence of a closed loop in which A_i acts n times :

$$L^{(n)} = \frac{(-ie_a)^n}{n} \int \ldots \int T_r\{K_+[n;1] \not A (1)K_+[1;2] \not A (2)\ldots K[n-1;n] \not A (n)\}d\tau_1 \ldots d\tau_n.$$

$$(12.12)$$

It can be shown via Furry's theorem that $L^{(n)}$ vanishes for odd n [cf. Bjorken and Drell 1965, for example].

Can we use the above result in the action at a distance framework? Not directly, since Feynman used field theory to arrive at it, i.e., he had to use quantization of the fermion fields. However, it is possible to rederive the result without recourse to second quantization, as was shown by Hoyle and Narlikar (1971). We briefly describe this work.

Like the propagator K_0 also define another K_0^- by the relation

$$K_0^-[2;1] = -\theta(t_1 - t_2)\sum_n u_n(2)\bar{u}_n(1). \qquad (12.13)$$

Thus the $K_0^-[2;1]$ propagator describes motion backwards in time within the past light cone at point 1. To have a more symmetric notation, we will denote $K_0[2;1]$ of Eq.(12.7) by $K_0^+[2;1]$. Both $K_0^\pm$ satisfy the inhomogeneous Dirac equation (12.7). $K_0^+[2;1]$ is non-zero along the future light cone at point 1.

Corresponding to these two propagators, we also distinguish between two types of paths, Γ_{21}^+ going forward in time and Γ_{21}^- going backwards in time in going from point 1 to point 2. Now use formulae (12.4) to define the amplitude $P(\Gamma_{21}^+)$ along Γ_{21}^+ while use a similar expression but with Γ_{21}^+ replaced by Γ_{21}^- to define the amplitude along Γ_{21}^-. The path integrals corresponding to these definitions ensure that

$$K_0^\pm[2;1] = \int P(\Gamma_{21}^\pm)\mathcal{D}\Gamma_{21}^\pm. \qquad (12.14)$$

Now suppose that we have a free particle in the four dimensional spacetime slab $t_1 \le t \le t_2$ and assume that the amplitude for the particle to reach $t = t_1$ from its previous history is given by $\psi_+(1)$. Likewise at $t = t_2$ we denote the amplitude for the particle to come from the future $t > t_2$ by $\psi_-(2)$. Fig. XII.2 illustrates this situation.

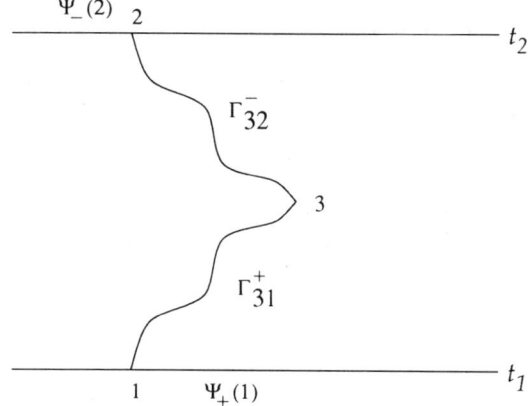

Fig. XII.2 In the slab contained within two time sections are wavefunctions $\psi_+(1)$ on $t = t_1$ and $\psi_-(2)$ on $t = t_2(> t_1)$. The probability amplitude at an intermediate piont 3 at $t = t_3(t_1 < t_3 < t_2)$ in the slab can be determined from the above boundary values. Typical forward going paths Γ_{31}^+ and backward going paths Γ_{32}^- convey this information from $\psi_+(1)$ and $\psi_-(2)$ respectively, each path having its own probability amplitude.

Thus the probability amplitude for the particle to be at an interior point 3 is given by $\psi(3)$ which is the sum of amplitudes along all Γ_{31}^+ type paths coming to it from points 1 on the time-section $t = t_1$ weighted by $\psi(1)$ together with amplitudes along all Γ_{32}^- paths from points 2 on the time section $t = t_2$ to the point 3 again weighted by $\psi(2)$:

$$\psi(3) = \int K_0^+[3;1]\gamma_4\psi_+(1)d^3\mathbf{x}_1 - \int K_0^-[3;2]\gamma_4\psi_-(2)d^3\mathbf{x}_2. \tag{12.15}$$

We impose as boundary condition that $\psi_+(1)$ is made of positive energy solutions while $\psi_-(2)$ is made of negative energy solutions. Then Eq.(12.15) is equivalent to the relation

$$\psi(3) = \int K_+[3;1]\gamma_4\psi_+(1)d^3\mathbf{x}_1 - \int K_+[3;2]\gamma_4\psi_-(2)d^3\mathbf{x}_2. \qquad (12.16)$$

Thus the Feynman propagator serves the convenient role of book-keeping of positive and negative energy states, of how the former travel forward in time along paths Γ^+ and the latter backwards in time along paths Γ^-.

Consider next the external potential A_i acting on the particle which is such that A_i vanishes outside the slab. In that case we may have scattering of paths backwards and forward in time due to the potential. Here again the K_+^A propagator helps in the book-keeping process. Only, we need to invoke the perturbation expansion to keep track of how many times the particle has been scattered.

To begin with, the amplitude for a path Γ_{21}^+ or Γ_{12}^- in the slab is defined by

$$P^A(\Gamma_{21}^+) = P(\Gamma_{21}^+) \exp\left[-ie \int_{\Gamma_{21}^+} A_i da^i\right]$$

$$\qquad (12.17)$$

$$P^A(\Gamma_{12}^-) = P(\Gamma_{12}^-) \exp\left[-ie \int_{\Gamma_{12}^-} A_i da^i\right]$$

with the understanding that point 1 is on $t = t_1$ and 2 on $t = t_2$. Paths within the slab need not be monotonic,however, with respect to t. Suppose we have a path from 1 to 2 with $2n$ reversals. Denoting intermediate points by i, sections i to $i + 1$ are monotonic, and the amplitude for such a path is given by

$$P^A(\Gamma_{21}) = \prod_i P^A(\Gamma_{i,i-1}^\pm) \qquad (12.18)$$

where the plus sign holds for the forward going sections and the minus sign for the backward going sections.

We return to a point that was taken for granted in the discussion so far. In the absence of an external potential A_i there would be no reversals, with the forward going paths continuing to go forward and likewise for the backward going paths. (Later we will reexamine this assumption in the light of the response of the universe; but for the time being we will continue with it.) In the presence of A_i, however, reversals can occur, with paths starting at t_1 ending also at t_1. In this case we get a ψ_- at $t = t_1$

even though there may be no ψ_+ at $t = t_2$. Similarly, we could have $\psi_+ \neq 0$ at t_2 originating solely from ψ_- at t_2. We therefore wish to know, what are ψ_+ on the time section $t = t_2$ and ψ_- on $t = t_1$ given a ψ_+ on $t = t_1$ and $\psi_- = 0$ on $t = t_2$?

Formally, we have

$$\psi_+^A(2) = \int \int P^A(\Gamma_{21})\gamma_4\psi_+(1)\mathcal{D}\Gamma_{21}d^3\mathbf{x}_1 \tag{12.19}$$

which includes paths with reversals. This expression can be calculated using the definitions given above and the standard path integral evaluation procedure [cf. Feynman and Hibbs, op.cit]. The calculation is given in Hoyle and Narlikar (1971) and we simply quote the result:

$$\psi_+^A(2) = \int K_+^A[2;1]\gamma_4\psi_+(1)d^3\mathbf{x}_1 \tag{12.20}$$

where,

$$
\begin{aligned}
K_+^A[2;1] = {}& K_+[2;1] - ie \int K_+[2;3]\,\mathbb{A}\,(3)K_+[3;1]d\tau_3 \\
& + (-ie)^2 \int \int K_+[2;4]\,\mathbb{A}\,(4)K_+[4;3]\,\mathbb{A}\,(3)K_+[3;1]d\tau_3 d\tau_4 + \ldots
\end{aligned} \tag{12.21}
$$

A similar analysis gives us $\psi^A(1')$:

$$\psi_-^A(1') = \int K_+^A[1';1]\gamma_4\psi_+(1)d^3\mathbf{x}_1, \tag{12.22}$$

where $K_+^A[1';1]$ is given by replacing point 2 in Eq.(12.21) by point $1'$ on the time section $t = t_1$.

Equation (12.21) is the perturbation expansion and is equivalent to Eq.(12.10). What about the factor C_v which appeared in the Feynman calculation ? Does it have an analogue in the particle sans field framework described here? We will consider this question next.

D. Vacuum Loops

The answer to the above questions is provided by including the as yet ignored feature of antisymmetrization of the wavefunction. The basic concept of indistinguishability of identical particles in quantum mechanics intervenes in the above analysis in the following way.

In Fig. XII.3(a) we have two particles going forward in time: the path Γ_{31}^{+} describes the motion of one particle from point 1 to point 3 while path Γ_{24}^{+} describes the motion of the other particle from point 4 to point 2. However, when we ask for the amplitude for there to be a particle at 3 and a particle at 2, given that we have a particle each at the earlier epochs at points 1 and 4, the answer must take note of the exclusion principle, and is obtained by subtracting from the amplitude for Fig. XII.3(a), the amplitude for Fig. XII.3(b) wherein we have interchanged the final states with the particle at point 1 going to point 2 and that at 4 going to 3.

What has been stated just now for a pair of particles also has relevance to the perturbation problem discussed above. In Fig. XII.3(c) we have a particle starting at point 1 and reaching a point 2 after being scattered twice by the potential A_i at points 3 and 4. Notice that the configuration of the Γ^{+} paths in this figure is the same as in Fig. XII.3(a) and so our antisymmetrization criterion leads us to include also the amplitude for the path configuration of Fig. XII.3(d) that corresponds to that of Fig. XII.3(b). This last diagram, however, describes a particle moving unperturbed from point 1 to point 2, along with a closed loop which has double scattering at points 3 and 4.

For a quantitative description of this idea see Hoyle and Narlikar (1971). The result can be stated as follows. If to zeroeth order the particle wavefunction at point 1 is $v_{+}(1)$ then to that order it is $v_{+}(2)$ at point 2. [The plus suffix denotes that it is made up of positive energy solutions.] Then corresponding to the Fig. XII.3(d), we have to add to the amplitude the term

$$-\frac{1}{2}(-ie)^2 \int \int T_r[K_+[3;4]\,\not{A}\,(4)K_+[4;3]\,\not{A}\,(3)]d\tau_3 d\tau_4.v_+(2) \qquad (12.23)$$

besides the second order term from the diagram XII.3(c) obtained earlier from the perturbation expansion (12.21). Notice that apart from the unity term, the coefficient of $v_+(2)$ in Eq. (12.23) is the lowest order term in the expansion of C_v. In fact, if we proceed further, using higher orders in the perturbation expansion following the antisymmetrization rule, we will recover the full factor C_v.

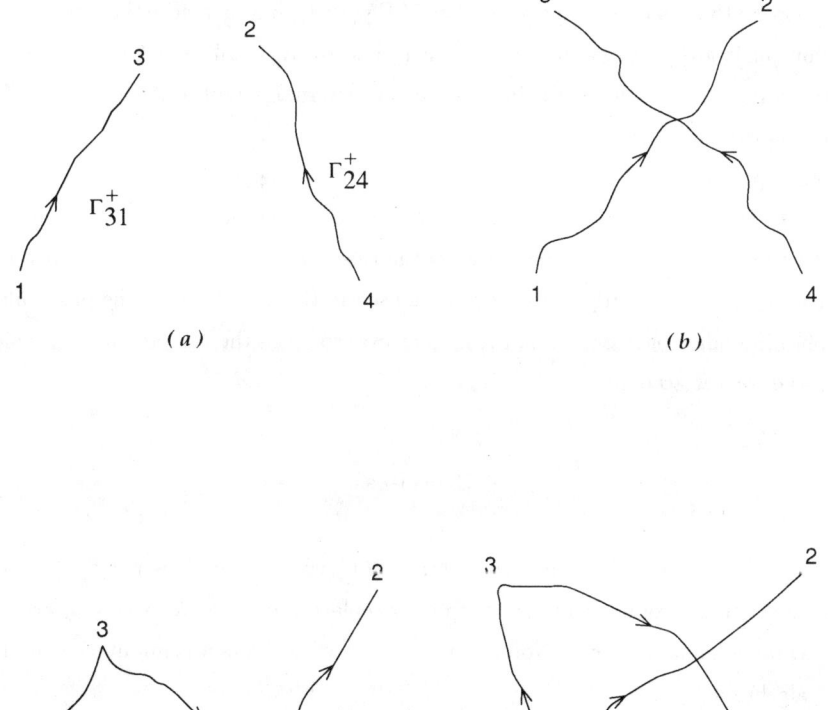

Fig. XII.3 As explained in the text the interchange of end-points of paths can lead to closed loops. The above example illustrates a typical case.

Hoyle and Narlikar (op.cit.) have also shown that the quantity L in Eq.(12.11) is equal to the path integral over the loop amplitudes defined as follows:

$$L = \int P(\Gamma^0) \exp \left[-ie \oint_{\Gamma^0} A_i dl^i\right] \mathcal{D}\Gamma^0 \qquad (12.24)$$

Here Γ^0 is a typical closed loop and the probability amplitude $P(\Gamma^0)$ is defined for a free loop by the same prescription as that of Eq.(12.4), but as before, the infinitesmal sections of it are propagated by $K_0^{\pm}$ rather than by K_+. Likewise the exponential factor in Eq.(12.24) denotes the influence of the external potential A_i on the loop Γ^0, following the same formula as Eq.(12.17).

The expression (12.12) can now be obtained from Eq.(12.24) by expanding the exponential phase factor and using a perturbation expansion. We leave it to the reader to demonstrate that this indeed is the case. In particular, it is easy to verify that even though the propagators $K_0^{\pm}$ were used in the definitions of the probability amplitude along a path or a loop, the final answer contains the K_+ propagators which keep the correct accounting.

Exercises

1. The objection against the use of Eq. (12.6) was that it does not incorporate spin. To get round it introduce spin by replacing da by $\gamma_i da^i$ where γ_i are the Dirac gamma matrices. Note, however, that with the non-commutative matrix algebra

$$\exp - \int_{\Gamma_a} im_a \gamma_i da^i \neq \prod_n \exp -im_a \gamma_i \Delta a_n^i$$

where the Δa_n^i are the segments of infinitensimal length into which the path Γ_a is divided. Which of the two expressions could qualify for a path integral formulation for a relativistic electron? Why?

2. A path is defined by a function $\mathbf{r}(t)$ in the range $0 \leq t \leq T$. If it represents a Dirac particle show that the probability amplitude function $P(t)$ for the path over the interval $[0, t]$ satisfies the differential equation

$$\frac{dP}{dt} = -im(\gamma_4 - \gamma.\dot{\mathbf{r}})P,$$

if the right hand side of the inequality of Q.1 is used. [for a further discussion of this problem see : Narlikar (1972).]

3. Consider the next higher order perturbation term than discussed in Fig. XII.2 to show how more vacuum loops can arise.

LECTURE XIII : MANY PARTICLE INTERACTIONS AND
THE QUANTUM RESPONSE OF THE UNIVERSE

A. The Problem of Many Particles

From the case of a single particle we now move on to the more realistic case of many particles. 'Realistic' because, as we have already noted, the notion of a single particle becomes a myth when we consider temporally backward and forward moving worldlines. We will also discover shortly that this aspect also seriously alters our classical notion of self action.

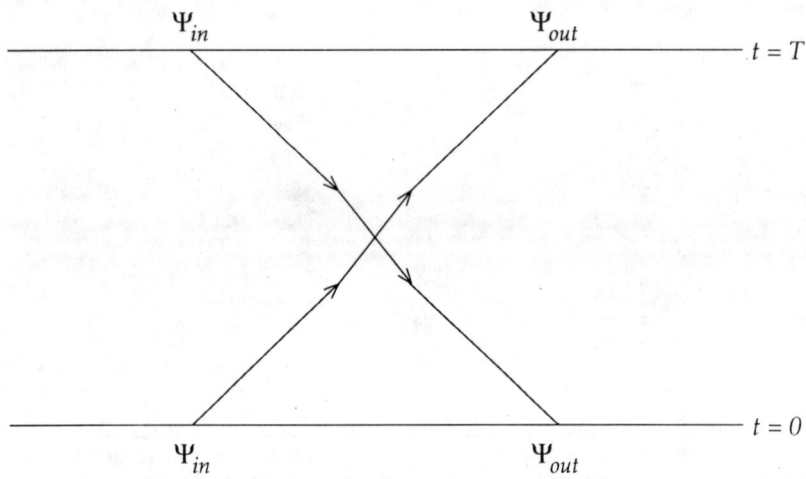

Fig. XIII.1 The surface S of the spacetime slab bounded by planes $t = 0, t = T$ has ψ_{in} defined on it in terms of wavefunctions of *electrons* at $t = 0$ propagating into $t > 0$ and *positrons* at $t = T$ propagating into $t < T$. Likewise, ψ_{out} denotes the wavefunction of outward bound particles, electrons on $t = T$ and positrons on $t = 0$.

We begin by restating the results obtained so far in a slightly different notation. We will denote our spacetime slab by $0 \leq t \leq T$ rather than by $t_1 \leq t \leq t_2$; and instead of denoting the wavefunctions by suffixes $\pm$ we will use the notation of "in" and "out" (see Fig. XIII.1). Thus ψ_{in} denotes what we earlier called ψ_+ on $t = 0$ and ψ_- on $t = T$, while ψ_{out} denotes what we earlier called ψ_+ on the time section $t = T$ and ψ_- on the time section $t = 0$. We will denote by S the surface of the slab and n^i will denote the unit outward normal to dS, the typical surface element.

To describe action at a distance we need to generalize Eq.(12.22) to a system of many particles $a, b, c, \ldots$ We will proceed step by step. First suppose that these particles are all in an external potential A_i but do not interact with one another. Then their wavefunction will follow the formula

$$\Psi_{\text{out}}[a', b', \ldots] = \int \int \ldots \int K_+^A[a'; a]\, \psi_a\, K_+^A[b'; b]\, \psi_b \ldots \Psi_{\text{in}}[a, b, \ldots] dS_a dS_b \ldots \quad (13.1)$$

Here the Ψ function has four spinorial components for each particle and the propagators act independently on these particles. From previous work we have

$$K_+^A[a'; a] = \int P(\Gamma_{a'a}) \{\exp [-ie_a \int_{\Gamma_{a'a}} A_i da^i]\} \mathcal{D}\Gamma_{a'a}, \quad (13.2)$$

where the typical path for particle a starts from point a and ends on point a' on the surface S. Thus the multi-particle propagator is simply the product of the individual propagators :

$$K[a', b', \ldots; a, b, \ldots] = \int \int \ldots \int P(\Gamma_{a'a})P(\Gamma_{b'b}) \ldots \exp [-ie_a \int_{\Gamma_{a'a}} A_i da^i - ie_b \int_{\Gamma_{b'b}} A_i db^i \ldots]$$
$$.\mathcal{D}\Gamma_{a'a}\mathcal{D}\Gamma_{b'b} \ldots$$
$$(13.3)$$

Stated in this form the transition to the interactive multi-particle system is natural. The formula (13.3) is generalized to

$$K[a', b', \ldots; a, b, \ldots] = \int \int \ldots \int P(\Gamma_{a'a})P(\Gamma_{b'b}) \ldots \exp [-ie_a \int_{\Gamma_{a'a}} A_i da^i - \ldots]$$
$$.\exp(iR)\mathcal{D}\Gamma_{a'a}\mathcal{D}\Gamma_{b'b} \ldots$$
$$(13.4)$$

where the extra factor $\exp iR$ in the path integral is none other than that given by the Fokker formula for interparticle action. We have

$$R = - \sum_{a<} \sum_b e_a e_b \int_{\Gamma_{a'a}} \int_{\Gamma_{b'b}} \delta(s_{AB}^2)\eta_{ik} da^i db^k. \tag{13.5}$$

We now show that all the well known results for quantum electrodynamics must follow from the formula (13.4) provided, as in the classical theory due note is taken of the response of the universe.

Before coming to grips with this fundamental problem we make one comment on the wavefunctions Ψ_{in} and Ψ_{out}. The exclusion principle requires the wavefunctions to be antisymmetric with respect to the interchange of any two particles, and if Ψ_{in} is antisymmetric then so is Ψ_{out}. But what does this mean in the present path integral approach? Note that the exclusion principle prevents two paths $\Gamma_{a'a}$, $\Gamma_{b'b}$ from crossing if $a \neq b$; more specifically, paths with common points make a zero contribution to the amplitude. Paths for the same particle can, however, cross since they are alternatives for the propagation of the particle. Thus we can tell from this property whether two paths belong to the same particle or different ones. Note also that the vacuum loops arose when we considered the antisymmetry property for a single particle.

B. The Influence Functional

We have already seen in the previous section how the quantum response of the universe appears in the form of an influence functional in the problem of transition of an atomic electron. We now look for a corresponding expression in the full relativistic problem of interacting electric charges.

To begin with, we note that in Eq.(13.4) it is the paths outside the slab, i.e., those satisfying the temporal conditions $t < 0$ or $t > T$ that contribute to the potentials A_i. The discussions of preceding sections tell us how to deal with the external potentials arising from the past portions, i.e., $t < 0$. The future portions contribute because of the $\delta(s^2)$ interaction. Without loss of generality we can take

$$A^i = A_{t<0}^i + A_{t>T}^i \quad ; \quad A_{t<0}^i = 0. \tag{13.6}$$

Then the future $(A_{t>T}^i)$ interactions in Eq.(13.4) are the so-called *response of the universe*.

Turning now to the interaction term R in Eq.(13.4), although we have written it in the classical fashion, we will later show that technically it should include the hitherto excluded self action terms $a = b$. For the time being we will continue with the classical expression (13.5) and exclude these terms.

Thus, we have,

$$-\sum_a \sum_{<b} e_a e_b \int \int \delta(s_{AB}^2)\eta_{ik}da^i db^k = -\frac{1}{2}\sum_a e_a \int A_{(a)i}da^i \qquad (13.7)$$

where,

$$A_{(a)i}(X) = \sum_{b \neq a} e_b \int \delta(s_{XB}^2)\eta_{ik}db^k = \frac{1}{2}[A_{(a)i}^{\text{ret}}(X) + A_{(a)i}^{\text{adv}}(X)]. \qquad (13.8)$$

Here we have separated the potentials into their advanced and retarded components as per our earlier discussion of classical direct particle electrodynamics. Therefore, the classical electrodynamic action is written in the form

$$-\sum_a e_a \int A_{it>T}da^i - \frac{1}{2}\sum_a e_a \int [A_{(a)i}^{\text{ret}} + A_{(a)i}^{\text{adv}}]da^i. \qquad (13.9)$$

(We have not considered the inertial part of the action here, which is of course assumed to be present.) If the universe is a perfect future absorber and an imperfect past absorber then the classical response is such that

$$A_i^R(X)_{t>T} = \frac{1}{2}\sum_a [A_i^{(a)}(X)^{\text{ret}} - A_i^{(a)}(X)^{\text{adv}}]. \qquad (13.10)$$

Hence the classical action becomes

$$-\sum_a e_a \int \{A_{(a)i}^{\text{ret}} + \frac{1}{2}[A_i^{(a)\ \text{ret}} - A_i^{(a)\ \text{adv}}]\}da^i. \qquad (13.11)$$

The first term in the above sum is the total retarded potential of all other particles $b \neq a$ while the second term is the Dirac radiation reaction formula. What is the corresponding quantum version of this result?

This is where we refer back to the discussion of Lectures IX–XI. There we saw that the apparently local behaviour of a quantum system actually involves the response of the universe via an influence functional which arises when we take into account how the absorber reacts back (via advanced potentials) on the local system. The influence functional enters into any probability calculation in the path integral approach (cf. Feynman and Hibbs, 1965) whenever the effects of external variables on the local system are integrated out. It is a double integral over paths and conjugate paths.

In Lectures IX–XI we saw how the conjugate paths arise in the calculation of probability for spontaneous transition of the atomic electron, involving the response of the universe, when the effects of the individual absorber particles are integrated out. The calculation requires besides paths $\mathbf{a}(t), \mathbf{b}(t), \ldots$ starting from points $a, b, \ldots$ etc. also the conjugate paths $\mathbf{a}'(t), \mathbf{b}'(t), \ldots$ which start from points $a^*, b^*, \ldots$ on S. But both the paths and conjugate paths end at the same points $a', b', \ldots$ respectively. As we saw in Lectures X and XI we end up with a transition probability instead of a transition amplitude. Experiments, in any case, are concerned with the measurements of the former only and so the theory does not suffer from any incompleteness on this count.

The paths and conjugate paths together permit the separation of positive and negative frequencies with the paths giving positive frequencies and the conjugate paths the negative frequencies. In the explicit example of Lecture X we saw how the quantum transitions in the future absorber lead to this distinction. We generalize the concept here.

Define the positive and negative frequency components of the advanced and retarded potentials by

$$A^i(X)^{\text{ret}}_{\pm} = \sum_b e_b \int \frac{\delta_{\pm}(t_X - t_B - |\mathbf{x} - \mathbf{x}_B|)}{2|\mathbf{x} - \mathbf{x}_B|} db^i, \tag{13.12}$$

$$A^i(X)^{\text{adv}}_{\pm} = \sum_b e_b \int \frac{\delta_{\pm}(t_X - t_B + |\mathbf{x} - \mathbf{x}_B|)}{2|\mathbf{x} - \mathbf{x}_B|} db^i \tag{13.13}$$

where the $\delta_{\pm}$ functions have the usual meaning

$$\delta_{\pm}(x) = \frac{1}{\pi} \int_0^{\infty} e^{\mp iwx} dw = \delta(x) \mp \frac{i}{\pi} \frac{\mathcal{P}}{x} \tag{13.14}$$

It can be shown that although the individual expressions defined in Eqs. (13.12) and (13.13) are not vectors, the differences

$$[A_+^{i\ \text{ret}} - A_+^{i\ \text{adv}}] \; ; \; [A_-^{i\ \text{ret}} - A_-^{i\ \text{adv}}]$$

transform as vectors. These combinations have no place in classical electrodynamics but they arise in quantum electrodynamics in a natural way.

The quantum response corresponding to the classical expression (13.10) is then given by

$$A_i^R(X)_{t>T} = \frac{1}{2}\sum_b \{[A_i^{(b)}(X)_+^{\text{ret}} - A_i^{(b)}(X)_+^{\text{adv}}] + [A_i'^{(b)}(X)_-^{\text{ret}} - A_i'^{(b)}(X)_-^{\text{adv}}]. \quad (13.15)$$

As stated above, the paths carry positive frequencies and the conjugate paths carry negative frequencies. If we imagine a coelescence of paths and conjugate paths Eq.(13.15) collapses into Eq.(13.10). To obtain an equivalent condition for $A_i'^R(X)_{t>T}$ we interchange paths with conjugate paths and positive with negative frequencies: and we find that

$$A_i'^R(X)_{t>T} = A_i^R(X)_{t>T}. \quad (13.16)$$

Thus (13.15) is symmetric with respect to paths and conjugate paths.

Typically, in the calculation of the influence functional we have two pairs of paths and conjugate paths $(\mathbf{a}, \mathbf{a}')$ and $(\mathbf{b}, \mathbf{b}')$ for pairs of interacting particles a, b. We therefore have four possible combinations. Consider the combination $(\mathbf{a}, \mathbf{b})$ which has two terms. The first is

$$-e_a e_b \int\int \delta(s_{AB}^2)\eta_{ik}da^i db^k$$

from the time symmetric interparticle action, and the second

$$-\frac{1}{2}e_a e_b \{ \int\int_{t_A > t_B} [\delta_+ (s_{AB}^2) - \delta_-(s_{AB}^2)]\eta_{ik}da^i db^k$$

$$+ \int\int_{t_B > t_A} [\delta_+(s_{BA}^2) - \delta_-(s_{BA}^2)]\eta_{ik}da^i db^k \}$$

arises from A_i. The time inequalities in the integrals reflect the advanced/retarded nature of the potential components. A little book-keeping exercise gives (cf. Hoyle and Narlikar 1974) the sum of the above two contributions as

$$-e_a e_b \int\int \delta_+(s^2_{AB})\eta_{ik}da^i db^k. \qquad (13.17)$$

Likewise, the combination $(\mathbf{a'}, \mathbf{b'})$ gives

$$+e_a e_b \int\int \delta_-(s^2_{A'B'})\eta_{ik}da'^i db'^k. \qquad (13.18)$$

The remaining two combinations $(\mathbf{a'}, \mathbf{b})$, $(\mathbf{a}, \mathbf{b'})$ combine to give

$$e_a e_b \left\{ \int\limits_{t_{A'}>t_B}\int \delta_+(s^2_{A'B})\eta_{ik}da'^i db^k - \int\limits_{t_B>t_{A'}}\int \delta_-(s^2_{BA'})\eta_{ik}da'^i db^k \right.$$

$$\left. + \int\limits_{t_{B'}>t_A}\int \delta_+(s^2_{B'A})\eta_{ik}da^i db'^k - \int\limits_{t_A>t_{B'}}\int \delta_-(s^2_{AB'})\eta_{ik}da^i db'^k \right\} \qquad (13.19)$$

We now have the influence functional in the form that generalizes Eq.(11.9) of Lecture XI :

$$F[\mathbf{a}, \mathbf{b}; \mathbf{a'}, \mathbf{b'}] = \exp\left[(e_a e_b/4\pi^2) \int d\Omega \int K dK. \right.$$

$$\left\{ \int\int \exp[-iK|t_A - t_B| + i\mathbf{k}.(\mathbf{x}_B - \mathbf{x}_A)]\eta_{ik}da^i db^k \right.$$

$$+ \int\int \exp[iK|t_{A'} - t_{B'}| + i\mathbf{k}.(\mathbf{x}_{A'} - \mathbf{x}_{B'})]\eta_{ik}da'^i db'^k \qquad (13.20)$$

$$- \int\int \exp[iK(t_A - t_{B'}) + i\mathbf{k}.(\mathbf{x}_{B'} - \mathbf{x}_A)]\eta_{ik}da^i db'^k$$

$$\left. \left. - \int\int \exp[ik(t_B - t_{A'}) + i\mathbf{k}.(\mathbf{x}_{A'} - \mathbf{x}_B)]\eta_{ik}da'^i db^k \right\} \right]$$

We interpret the various terms in $F[\mathbf{a}, \mathbf{b}; \mathbf{a'}, \mathbf{b'}]$ as follows. For $t_A > t_B$, the positive terms in the curly brackets contribute to a downward transition of particle b and an upward transition of particle a, and vice versa for $t_B > t_A$. The negative terms contribute to downward transitions of both a and b. Thus, (13.20) describes absorption and stimulated emission. In general we may consider these phenomena as energy exchanges between particles of a pair and between the pair and the surroundings. Since the coefficient in front of all terms is the same, the probabilities for these processes are also the same.

Exercises

1. Discuss the role of paths and conjugate paths in the way the absorber responds to a local signal.

2. Discuss why the response function (13.15) has to be symmetric with respect to a pair of path and conjugate path.

3. Consider each exponential in Eq. (13.20) to verify the stated role of that term in quantum electrodynamics, in the text that follows it.

LECTURE XIV : SELF ACTION

A. A Self-interacting Charge?

We now come to an issue that distinguishes the quantum treatment of direct inter-particle action from its classical treatment. Suppose self action were included in the classical theory. This would mean the addition of an integral of the following form for each typical charge a in the action formula (1.6):

$$-\frac{1}{2}e_a^2 \int \int \delta(s_{A\tilde{A}}^2)\eta_{ik}da^i d\tilde{a}^k. \tag{14.1}$$

Here both points A and $\tilde{A}$ lie on the same path $a(t)$. Since all classical paths are timelike, we have $\delta(s_{A\tilde{A}}^2) = 0$ for $A \neq \tilde{A}$. That is, for two distinct points on the path the segment $A\tilde{A}$ is time-like and so the deltafunction vanishes. Hence the term adds nothing to the action except in the case $A = \tilde{A}$. It turns out, however, that this addition leads to the notorious infinities of electrodynamics. This is therefore absent in the Fokker action.

In quantum electrodynamics, the situation is different since the paths here are made up of both Γ^+ and Γ^- segments,i.e., they can turn backwards in time. Thus we can find distinct points A and $\tilde{A}$ on the same path $a(t)$ such that $s_{A\tilde{A}}^2 = 0$. In other words segments on which these points lie do interact. Can we therefore retain the rule that there is no self interaction of a typical path? This would mean that two segments with points on one connectible to points on the other by null rays still do not interact if they belong to the same path.

Such a rule would be difficult to implement in practice since it would require know-ing beforehand the full history of the path or paths to which the segments belong in order to decide whether they interact or not. Besides, experimental results don't support such a rule. Consider, for example, the case of the unstable positronium.

A positronium is made of a positron-electron pair going round each other as in a hy-drogen atom but with the proton replaced by the positron. The fact that the positron interacts by a Coulomb force with the electron tells us in the action at a distance format that the two are 'different' particles. Yet after a short while the positron and electron annihilate each other providing energy. With the usual interpretation of the phenomenon, it means that the worldlines of the positron and the electron are the Γ^+ and Γ^- sections of the worldline of a single charge. Thus the Coulomb action is of the charge on itself. To avoid such a conclusion we have to consider another possibility.

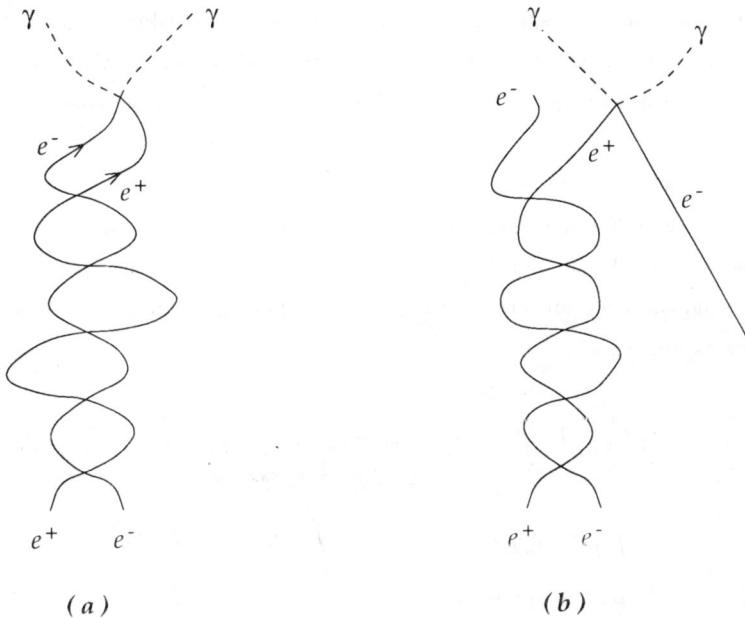

Fig. XIV.1 In (a) we have the worldlines of the positron and electron in the *positronium* atom annihilating each other. In (b), the e^+ part annihilates against a neighbouring electron e^-. If (a) were forbidden then the rate of annihilation should drop to zero as the density of the medium is reduced to near vacuum. In reality, the rate does not drop below a threshold that corresponds to (a).

Could it be that the positronium annihilation takes place via an alternative route? This route would involve the annihilation of the positron not by the electron of the positronium but by an ambient electron. This is of course, possible. However, if self annihilation of the positronium were to be ruled out, then the annihilation rate would depend on the density of the surrounding medium. If the medium is gradually evacuated then the rate of annihilation should drop to zero. See Fig. XIV.1 for details.

Experiments, however, show that the rate drops to a finite value which corresponds to self annihilation. So we are driven back to the first conclusion that the positron and electron can form part of the worldline of the *same* electric charge. Thus it seems necessary to add the self action term (14.1) to the action and limit the lack of self-interaction to the proviso $A \neq \tilde{A}$.

Including the self action therefore,we need to consider the influence functional for the case $a = b$. Hoyle and Narlikar (1971) have shown that in such a situation the essential contributions are from the combinations $(\mathbf{a}, \mathbf{a}'), (\mathbf{a}, \mathbf{a})$ and $(\mathbf{a}', \mathbf{a}')$. The resulting expression is

$$
F[\mathbf{a}, \mathbf{a}'] = \exp\{ie_a^2[\int\limits_{t_{A'} > t_A}\int \delta_+(s_{A'A}^2)\eta_{ik}da'^i da^k - \int\limits_{t_A > t_{A'}}\int \delta_-(s_{AA'}^2)\eta_{ik}da'^i da^k
$$

$$
-\frac{1}{2}\int\int \delta_+(s_{A\tilde{A}}^2)\eta_{ik}da^i d\tilde{a}^k + \frac{1}{2}\int\int \delta_-(s_{A'\tilde{A}'}^2)\eta_{ik}da'^i d\tilde{a}'^k]\}
$$

(14.2)

or, in terms of Fourier integrals it becomes

$$
F[\mathbf{a}, \mathbf{a}'] = \exp\Big[(e_a^2/4\pi^2)\int d\Omega \int\limits_0^\infty K dK\{-\int\int \exp[iK(t_A - t_{A'}) + i\mathbf{k}.(\mathbf{x}_{A'} - \mathbf{x}_A)]\eta_{ik}da^i da'^k
$$

$$
+ \int\limits_{t_A > t_{\tilde{A}}}\int \exp[iK(t_{\tilde{A}} - t_A) + i\mathbf{k}.(\mathbf{x}_{\tilde{A}} - \mathbf{x}_A).\eta_{ik}da^i d\tilde{a}^k
$$

$$
+ \int\limits_{t_{A'} > t_{\tilde{A}}}\int \exp[iK(t'_{A'} - t_{\tilde{A}'}) + i\mathbf{k}.(\mathbf{x}_{\tilde{A}'} - \mathbf{x}_{A'})]\eta_{ik}da'^i d\tilde{a}'^k\}\Big].
$$

(14.3)

This is the same result as that obtained in Lecture XI. The first term in the curly bracket gives the spontaneous transitions while the second and third terms contribute the radiative correction effects. In principle a factor of this form should exist for all particles $a, b, c, \ldots$ of the system. We will consider the radiative corrections in the next lecture.

Here we do emphasize, however, that the inclusion of (14.1) brings us on par with quantum field theory with its advantages as well as disadvantages. The former are the ability of the theory to predict (correctly) the Lamb shift, the anomalous magnetic moment of the electron and other similar effects. The disadvantages are that unless we find a logical way of preventing A and $\tilde{A}$ from coinciding in (14.1) we land up with the usual infinities. In quantum field theory formal but artificial renormalization techniques help prevent A and $\tilde{A}$ from coinciding. The same techniques could be invoked in the present theory, but as we shall see in the next lecture, there is a natural cut off introduced by cosmology that does indeed prevent A and $\tilde{A}$ from coinciding.

B. Interaction with Vacuum Loops

The espression (13.4) is still incomplete because we have not included the interactions of paths with loops. As discussed in Lecture XII. the loops arise in the theory by the requirement of antisymmetrization and thus will influence any phenomenon of interaction between "real" particles.

This generalization is straightforward, given the earlier discussion of the influence functional for paths of particles. As shown by Hoyle and Narlikar (1971) the loops by themselves do not affect the probability calculation but the interactive term between a loop and a particle path does. The influence functional for a loop-particle interaction is given by

$$
\begin{aligned}
F[\mathbf{a}, \mathbf{l}; \mathbf{a}', \mathbf{l}'] = \exp[iee_a\{ &\int\limits_{t_{A'} > t_L} \int \delta_+(s^2_{A'L})\eta_{ik}da'^i dl^k - \int\limits_{t_L > t_{A'}} \int \delta_-(s^2_{LA'})\eta_{ik}da'^i dl^k \\
&+ \int\limits_{t_{L'} > t_A} \int \delta_+(s^2_{L'A})\eta_{ik}da^i dl'^k - \int\limits_{t_A > t_{L'}} \int \delta_-(s^2_{AL'})\eta_{ik}da^i dl'^k \\
&- \int \int \delta_+(s^2_{AL})\eta_{ik}da^i dl^k + \int \int \delta_-(s^2_{A'L'})\eta_{ik}da'^i dl'^k \}].
\end{aligned}
$$

$$(14.4)$$

As we shall see, this type of interaction produces the effects normally ascribed to "vacuum polarisation" in field theory.

Here again, the exact similarity with quantum field theory brings its difficulties of divergent integrals. However, in the next lecture we will show how to deal with such matters in a consistent way.

With this result, formally at least, the action at a distance quantum electrodynamics can be said to have reached the same level of attainment as the conventional quantum field theoretic electrodynamics. Although we have used (throughout this section) the language of flat spacetime, we have done so because electrodynamics is conformally invariant and our cosmological models are conformally flat. Thus our arguments work in an expanding universe with the correct past and future boundary conditions, i.e. with the response required by the condition (13.15). However, it is cosmology that brings out the real differences between the two approaches when we consider the so called radiative corrections and the renormalization programme. We turn to these issues next.

Exercises

1. Relate the length scale separating points A and $\tilde{A}$ in the self action integral (14.1) to a frequency and comment on the so called ultra-violet divergence as a consequence of $\tilde{A} \to A$.

2. Investigate the positronium annihilation rate quantitatively from standard textbooks on quantum electrodynamics and arrive at the figure XIV.1(a).

3. Explain why the classical picture of excluding self action does not work in quantum electrodynamics.

LECTURE XV : COSMOLOGICAL CUT-OFFS
TO RADIATIVE CORRECTIONS

A. Radiative Corrections

We now examine the so-called self energy correction due to the radiative processes in action at a distance electrodynamics. Recall that the classical self energy problem is solved in this theory by the use of advanced reaction from the rest of the universe. The problem arises in quantum field theory from the ultraviolet divergence, i.e., from the degrees of freedom of the electromagnetic field of very high frequency. In the action at a distance version, the self energy problem in principle appears from the identification of the two legs $A, \tilde{A}$, of the deltafunction in the interaction

$$-\frac{1}{2}e_a^2 \int \int \delta_+(s_{A\tilde{A}}^2)\eta_{ik}da^i d\tilde{a}^k \qquad (15.1)$$

describing the action of a charge a on itself. Note that the deltafunction δ_+ instead of δ appears in Eq.(15.1) after we include the response of the universe.

Any computation of a quantum electrodynamic cross section using Eq.(15.1) leads to a divergent result if one uses it literally as given. However, action at a distance requires a lower cut off of a kind

$$|X_A^i - X_{\tilde{A}}^i| \geq |\epsilon^i| \qquad (15.2)$$

with the cut off vector ϵ^i having length ϵ small compared to the Compton wavelength of the mass m_a. Hoyle and Narlikar (1971) had conjectured that if a more complete theory includes classical gravity then a natural cut off would be the Schwarzschild radius of the charge, $2Gm_a/c^2$. Later Padmanabhan (1985) showed that in a quantum gravity context the cut off is the Planck length $(G\hbar/c^3)^{\frac{1}{2}}$.

Neither of these cut offs, however, reflect the global nature of the problem, i.e., the fact that any local quantum measurement is subject to the interference of the response of the universe. A clue to this type of cut off was provided by our earlier discussion of Lecture XI. There we found that because of the event horizon in the future absorber the response is limited to frequencies upto those satisfying the inequality

$$k < \omega_{\text{eff}}/HT. \tag{15.3}$$

whre T is the time duration of the measurement. This was the limit in the spontaneous transition problem. But, as we saw in Lectures XIII and XIV, the limit will appear in the more general influence functional calculated in formulae (13.20) and (14.3). This limit to high frequencies in the momentum space translates to a lower limit in the configuration space. Identifying the lower limit ϵ in formula (15.2) with that in Eq. (15.3) we get

$$\epsilon \sim k^{-1}. \tag{15.4}$$

With a finite cut-off the calculation of the bare and observed masses of the electric charge can be performed using the usual methods of the renormalization programme. [See Hoyle and Narlikar 1971 for details.]

The cut off on k at the high frequency and given by (15.3) works out to $\sim 10^{31} s^{-1}$ for the atomic and cosmological parameters described above. This cut off may vary from one microscopic process to another; it also is linked with the properties of the cosmic absorber. However, the reasoning given above tells us that *for every microscopic process in electrodynamics a cut off exists.*

The purely local approach to QED demands Lorentz invariance in every operation that may be performed. Our method, on the other hand, picks out a specific local reference frame, viz. the so called *cosmological rest frame*, to define the response of the universe. Thus Lorentz invariance is manifestly not present, although one can *use* the Lorentz transformation to describe any process of QED in a frame different from the cosmological rest frame.

Choose the cosmological rest frame in which, by the arguments of the preceding section all Fourier integrals in the computations of the influence functional have a high frequency cut off at k_{max}, say. With $c = 1$, $\hbar = 1$, this cut off corresponds to a restriction in time coordinate

$$|t_P - t_Q| \geq k_{max}^{-1}. \tag{15.5}$$

We will shortly specify k_{max}; for the time being k_{max}^{-1} remains a small quantity akin to ϵ used in Eq. (15.2).

Using the derivation of Hoyle and Narlikar [1971 : see their Eq. (146)] therein we find that *in the cosmological rest frame* the observed mass m_{obs} is related to the theoretical mass m_{th} of the electron by the relation

$$m_{obs} = m_{th}\left\{1 + \frac{3e^2}{2\pi}ln\left(\frac{k_{max}}{m_{th}}\right)\right\}. \tag{15.6}$$

Notice that a similar formula comes from quantum field theory but there the cut off is a purely abstract quantity and so no numerical significance is attached to the mass difference

$$\Delta m = m_{obs} - m_{th}. \tag{15.7}$$

In the present theory k_{max} is related to physical parameters and as a result it is possible to estimate it and Δm, which we proceed to do now.

The upper limit on k is given by (15.3) in which we have $\omega_{eff} \sim 80s^{-1}$, $H^{-1} \sim 3 \times 10^{17}s$ and T a time scale large compared to the characteristic time for the process; in this case the free motion of the electron. We may take $T \sim \hbar/mc^2 \sim 10^{-21}s$. Thus we have

$$k_{max} \sim 10^{40}s^{-1}. \tag{15.8}$$

Using these values, (15.6) and (15.7) give in dimensionless form

$$\frac{\Delta m}{m} \sim \frac{3e^2}{2\pi\hbar c}ln(10^{19}) \sim 0.15. \tag{15.9}$$

In fact, with $T \sim \hbar/mc^2$, (15.9) expressed in symbols is

$$\frac{\Delta m}{m} \sim \frac{3\alpha}{2\pi}ln\left(\frac{\omega_{eff}}{H}\right), \tag{15.10}$$

where α is the fine structure constant.

Two comments are needed to elaborate the above conclusion. First, (15.10) shows clearly the cosmological input to the correction term which no purely local attempt at resolving the divergence problem will arrive at. Second, the correction has been obtained in the cosmological rest frame : and so the statement is not strictly Lorentz invariant. This in our view is an unavoidable conclusion echoing the first comment that only a global theory can lead to the resolution of the divergence problem. In this context we again notice Dirac's intuitive perception when he wrote :

> "With a cut-off we eliminate at once all difficulties about divergent integrals which have been plaguing theoretical physics for decades. These difficulties arise only because people want to have strict Lorentz invariance in an imperfact theory. In doing so they are aiming for something which may very well be impossible". [- Dirac 1969]

We would agree with the above sentiment with one modification : replace the adjective 'imperfect' by 'incomplete' to underscore the one crucial element missed out in field theory, viz. the response of the universe.

B. Charge Renormalization

The "renormalization" of electric charge occurs in the present theory through the interaction of charged closed loops as intermediaries between the interaction of any two charges. Fig.(XV.1) for example illustrates the lowest order such process. The loop-path influence functional then leads to the amplitude

$$-(-ie^2)(-ie^2)\int\int\int \bar{u}_2(1)\gamma_i u_1(1)\delta_+(s_{12}^2)Tr[\gamma^i K_+(2;3) \not{B} (3)K_+(3;2)]d\tau_1 d\tau_2 d\tau_3$$

$$(15.11)$$

with the integrations with respect to τ_1, τ_2 and τ_3 being over the slab $0 \leq t \leq T$. The scattering is the same as that produced by a potential

$$A^i(1) = ie^2 \int\int \delta_+(s_{12}^2)Tr[\gamma^i K_+(2;3) \not{B} (3)K_+(3;2)]d\tau_2 d\tau_3, \qquad (15.12)$$

which in turn corresponds to the current

$$j^i(1) = \frac{1}{4\pi}\Box_1 A^i(1) = ie^2 \int Tr[\gamma^i K_+(1;3) \not{B} (3)K_+(3;1)]d\tau_3 \qquad (15.13)$$

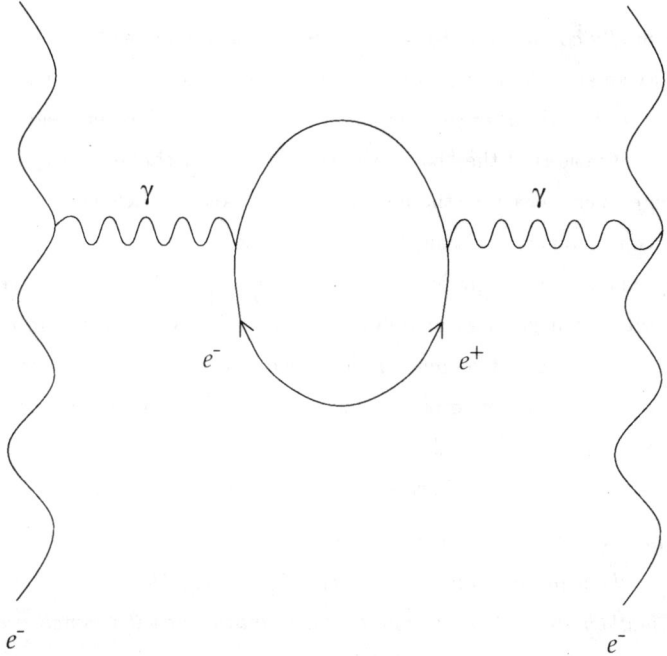

Fig. XV.1 The lowest order process involving a closed loop in the vacuum as an intermediary in the interaction between two charges.

The evaluation of $j^i(1)$ is given in detail by Hoyle and Narlikar(1971) and we simply quote the result which is cut off dependent :

$$j_i(1) = \frac{2e^2}{3\pi} ln(m\epsilon) J_i(1) \qquad (15.14)$$

where m is the mass of the charge. The ϵ here arises again from the lower limit on the separation of the two legs of the deltafunction and its interpretation in terms of the upper limit on frequencies of the absorber response is as given by Eqs (15.3) and (15.4). Thus in terms of the cosmological parameters of the absorber we get the charge modification Δe as

$$\frac{\Delta e}{e} = \frac{2e^2}{3\pi\hbar c} ln\left(\frac{\omega_{\text{eff}}}{H}\right). \qquad (15.15)$$

The result is that closed loops effectively lower the theoretical (or "bare") value of the electric charge by some 0.04 fraction of its original value.

We may briefly comment on the relationship (15.4) further as follows. For A and $\tilde{A}$ separated by smaller than ϵ the corresponding frequencies in the influence functional are too high for the absorber to react and influence the local experiment. This results in slight modifications of the "bare" values of mass and charge of a typical particle; for, any experiment measures these values not for isolated (i.e. bare) charges but for charges in continuous interaction with the universe.

The renormalization programme in the quantum field theory of charged particles has the merit that it gives an unambiguous way of handling infinite integrals which are only logarithmically divergent. It has been felt that the actual values of these integrals do not contribute to observable quantities and as such the success of the programme is judged by how the residuals left after removing the infinite integrals pass the observational tests. However, as Dirac observed :

> "... this so called good theory (QED) ... involves neglecting infinities, ne-
> glecting them in an arbitrary way. This is not sensible mathematics. Sen-
> sible mathematics involves neglecting a quantity when it is small—not ne-
> glecting it just because it is infinitely great and you do not want it." [– Dirac
> 1978]

The proposed remedy in the present approach solves this outstanding difficulty—at a price that the theoretical physicist trained at viewing the problem in a purely local way will find it difficult to appreciate . Yet, the merit of the solution presented here should induce him to take into account the missing link, namely the response of the universe. It is this link that forces us to consider cosmological boundary conditions for seemingly local problems.

Exercises

1. Vary the cosmological parameters H and ω_{eff} by a few powers of 10 to see how sensitive the correction $\Delta m/m$ is for these variations.

2. Repeat the above exercise for $\Delta e/e$.

3. Because the radiative correction singles out a specific Lorentz frame is it possible to think of an experiment in quantum electrodynamics which will measure
   motion of the laboratory rest frame relative to the cosmological rest frame?

LECTURE XVI : CONCLUDING REMARKS

A. Experimental Search for Advanced Potentials

Assuming the above approach to electrodynamics to be valid, it follows that the nature of the accepted cosmological model should be consistent with the local experiments of electrodynamics. In particular, if the cosmological response is not such as to give pure retarded solutions, then it may be possible to detect advanced effects. There have been attempts to look for small advanced effects in local radiation experiments, although their interpretation itself is shrouded in controversy.

As we saw in earlier sections, no standard big bang cosmology satisfies the absorber condition to give unambiguous pure retarded solutions. It follows therefore that if one of these cosmologies is right then the pure retarded solution is untenable. Can it be that the incompleteness of future absorption would show itself through the presence of small fractions of advanced effects in local experiments?

Partridge (1973) attempted to detect such an effect in the radiation of a microwave source as it alternately radiated into free space and a local absorber. Partridge argued that advanced potentials lead to power gain rather than power loss in the source. Hence if a tiny fraction of radiation is via advanced potentials, the power drain from the source would be less than in the pure retarded case.

Partridge set up an arrangement in which radiation was blocked by a local absorber in one direction and was allowed to move freely in another. The argument was that the local absorber will ensure pure retarded effects whereas the radiation into free space would travel long distances and through an incompletely absorbing universe. A switching arrangement allowed these possibilities alternately. Within the accuracy of the experiment (estimated at 1 part in 10^8) there was no difference in the two cases. Thus Partridge claimed to have found no evidence of advanced effects.

Subsequently, Heron and Pegg (1974) argued that Partridge's use of a static absorber would inevitably lead to a null result. Instead what they proposed was an experiment with a time asymmetric *chopper absorber* to alter the boundary conditions. This would allow them to alter the ratio of advanced to retarded components, leading to a possible detection of the former.

However, Davies (1975) has criticised both the above approaches on the grounds that with proper inclusion of thermodynamics, attempts like these are bound to give null results. The objection raised by Davies goes in fact deeper than the specific issues relating to the proposed experiments. Davies has argued that one cannot bypass thermodynamics as proposed by Hogarth, Hoyle and Narlikar and that ultimately the thermodynamic asymmetry like that in the Boltzmann H-theorem will have to be included in any realistic discussion of electrodynamic time asymmetry. In other words, Davies was reverting to the explanation of time asymmetry given by Wheeler and Feynman (1945) referred to in Lecture III.

While this could be a possible line of argument it misses the entire spirit of the action at a distance theory. First, it postulates *ad hoc* asymmetrical initial conditions which are basic to the H-theorem. Secondly, once one decides to work within the action at a distance framework the non-locality of the problem forces one to take cognizance of the large scale structure of the universe, and the cosmological considerations become relevent and unavoidable. The self-consistent mixture of advanced and retarded potentials is determined by including the response of the universe. Finally, rather than use the statistical laws of thermodynamics as fundamental laws, attempt should be made to understand them as a consequence of other more fundamental arrows of time like electrodynamics and cosmology.

B. Relationship to Other Arrows of Time

Another aspect of the Partridge type experiment relates to the deeper question of a relationship of thermodynamic and electrodynamic arrows to the expanding and contracting phases of a time-symmetric universe, such as the Friedmann model with $k = +1$. Do these time arrows reverse when the universe contracts? This question, so long as one sticks to the Wheeler-Feynman electrodynamics, is not uniquely answered. Recently Gell-Mann and Hartle (1991) have discussed this problem in a different way. They investigate a way in which the rules of quantum mechanics might be adapted to impose a time symmetry on the boundary conditions. Thus when the universe enters the contracting phase, these microscopic degrees of freedom of the universe conspire to reverse the time-asymmetric processes. In a reanalysis of the Partridge experiment Davies and Twamley (1993) argue that its null result goes against the Gell-Mann Hartle model, but suggest that a more stringent test would be to repeat the Partridge experiment with a laser rather than microwave antenna. This is because the universe is apparently transparent out to great distances at the GHz frequencies and to include its absorptive effects along the future light cone in a more significant way much higher frequencies should be used.

The above remarks motivate important extension of action at a distance concept to all basic interactions of physics, so that the thermodynamics of macroscopic systems can be understood as a consequence of the largest scale time asymmetry, viz.,the expansion of the universe. Thus one needs to go deeper into the controversial issue (see Gold, 1968 for a discussion) as to whether in a contracting universe thermodynamics goes in the reverse direction. Thus if thermodynamic and biological clocks are reversed in a contracting universe, we humans should continue to see an expanding universe as we get older!

C. Action at a Distance as a Universal Phenomenon?

The above ideas take us beyond electrodynamics which has been the main topic of this book. There are other compelling reasons for seeking such an extension, for, so far as electrodynamics is concerned action at a distance has now demonstrated the following advantages over the field theoretic description :

1. The choice of retarded solutions is not *ad hoc* as in field theory but dictated by the time asymmetry of the universe.

2. There is no paradox involving infinities due to self action in the classical theory.

3. The theory is able to account for all classical as well as quantum electrodynamics without the extra degrees of freedom vested in free fields, i.e., it is more economical in its postulates.

4. The cosmological boundary conditions provide the cut off at high frequencies and thereby eliminate the divergences that normally plague quantum electrodynamics.

5. The concept of the response of the universe provides a powerful tool for limiting viable cosmological models.

These advantages are sufficient to motivate the generalization of action at a distance to other areas of physics. Hoyle and Narlikar (1964, 1974) have shown how this can be done for gravity. There the starting point is inertia of matter, with mass defined as a direct particle field with origin in matter. Just as with less than two charges there is no electrodynamics, for less than two particles there is no inertia and no gravity. This approach leads to a remarkable synthesis of Mach's principle with general relativity. The theory reduces in the many particle approximation to general relativity with the additional demonstration that the sign of the gravitational constant has to be positive.

More recently, Hoyle et al(1994b) have further generalized the formulation to describe the creation of matter, including the deduction of the cosmological constant. Here the creation is through the basic unit of Planck mass $(\hbar c/G)^{1/2}$ which subsequently decays through a series of high energy physics interactions to baryonic matter... a process yet to be determined by the particle physicists. This may very well involve a "grand unification" but through action at a distance instead of fields.

What have been other attempts towards extending the action at a distance formulation to other interactions? In the late sixties Narlikar (1969) showed how to construct an action at a distance counterpart for a field theory of arbitrary spin having a quadratic Lagrangian and linear field equations. Such a formulation will naturally have an "absorber theory" similar to the Wheeler-Feynman theory. Earlier Narlikar (1962) had discussed an absorber theory involving neutrinos on lines similar to virtual photons assuming that the neutrinos travel with the speed of light and mediate in weak interactions.

There is one further hint of the possible role of the response of the universe in local phenomena, a role that takes us to issues believed to be related to the very foundations of quantum mechanics. The discussions of Lectures IX–XI and XII–XIII tell us that it is not proper to talk of a probability amplitude for a local microscopic system. The correct description of the physical behaviour of the system follows from the probability calculation that includes the response of the universe. Thus one is dealing with a "square of the amplitude" type of expression rather than the amplitude itself.

This may explain the mystery that surrounds such epistemological issues like the *collapse of the wavefunction*, issues that are discussed entirely in a local framework. What is missing from the usual discussion of the problem is the response of the universe. The wavefunction collapse represents the final course of action taken by the system consistently with the response of the universe. We suggest this idea as a way of understanding many other conceptual issues of quantum mechanics. It may well be that the real non-local "hidden variables" are really global, being contained in the response of the universe. For a detailed discussion of this idea see Hoyle (1982) and Narlikar (1993b).

Perhaps the stiffest resistence to the concept of action at a distance would come today not from microphysics but from cosmology. As we found earlier, all the popular big bang models fail to meet the appropriate temporal boundary conditions while the steady state model satisfies them. However, the latter model has several difficulties in explaining the observed large scale features of the universe. [So does the big bang idea; but it has gained acceptance because of the belief that only it can provide an explanation for the microwave background and the abundances of light nuclei, which had baffled the steady state model.]

Recently, however, Hoyle et al(1993, 1994a) have produced the so-called "quasi-steady state cosmology" (QSSC) that appears to circumvent the difficulties faced by the steady state model. This model combines some features of the big bang model with some of the steady state model. It is thus able to explain observed features like the microwave background, abundances of light nuclei, the redshift magnitude relation for galaxies,radio source counts, angular size redshift relation, etc that are normally claimed as successes of the big bang cosmology. It also explains features which that cosmology finds hard to accommodate like the age distribution of galaxies, baryonic dark matter, relationship to high energy astrophysics,and above all the explanation of the primary creation of matter within a framework that respects the law of conservation of matter and energy.

Although not claimed as "the cosmology" by its authors, the QSSC therefore is indicative of the kind of cosmology that may be required to accommodate the growing list of extragalactic phenomena being discovered. For the present work, it has the merit of generating the right kind of cosmological response. We therefore end this book with the hope that in the growing interaction between fundamental physics and cosmology the action at a distance approach may have a lot to offer to both the disciplines.

REFERENCES

Bjorken, J.D. and Drell, S.D. 1965, Relativistic Quantum Fields, Mcgraw Hill, New York.

Bondi, H. and Gold, T. 1948, Mon. Not. R. Astron. Soc. **108**, 252.

Brans, C. and Dicke, R.H. 1961, Phys. Rev. **124**, 125.

DeWitt, B.S. and Brehme, R.W. 1961, Ann. Phys. (U.S.A.) **9**, 220.

Dirac, P.A.M. 1938a, Proc. R. Soc. **A165**, 199.

Dirac, P.A.M. 1938b, Proc. R. Soc. **A167**, 148.

Dirac, P.A.M. 1969 In *Fundamental Interactions at High Energy*, Eds T. Gudehus, G. Kaiser and A. Perlmutter (New York : Gordon and Breach).

Dirac, P.A.M. 1978, in *Directions in Physics*, 36, New York : Wiley.

Davies, P.C.W. 1972, J. Phys. A : Gen. Phys. **5**, 1722.

Davies, P.C.W. 1973, Mon. Not. R. Astron. Soc. **161**, 1.

Davies, P.C.W. 1975, J. Phys. A : Math. Gen. **8**, 272.

Davies, P.C.W. and Twamley, J. 1993 Class. Quantum Grav. **10**, 931.

Einstein, A. and deSitter, W. 1932, Proc. Nat. Acad. Sci. **18**, 213.

Feynman, R.P. 1949, Phys. Rev. **76**, 749

Feynman, R.P. 1950, Phys. Rev. **80**, 440

Feynman, R.P. 1963, Private Communication; but also see Mr. X in the *The Nature of Time*, Ed. T. Gold (Cornell, 1967).

Feynman, R.P. and Hibbs, A.R. 1965, *Quantum Mechanics and Path Integrals*, (McGraw Hill, New York).

Fokker, A.D. 1929 a, Z. Phys. **58**, 386.

Fokker, A.D. 1929 b, Physica **9**, 33.

Fokker, A.D. 1932, Physica **12**, 145.

Friedman, A. 1922, Z. Phys. **10**, 377.

Friedman, A. 1924, Z. Phys. **21**, 326.

Gauss, C.F. 1867, Werke **5**, 629.

Gell-Mann, M. and Hartle, J.B. 1991, Santa Barbara Preprint UCSBTH–91–31.

Gold, T. 1968 *The Nature of Time* (Cornell University Press).

Heron, M.L. and Pegg, D.T. 1974, J. Phys. A : Math., Nucl. Gen., 7, 1965.

Hogarth, J.E. 1962, Proc. R. Soc. **A314**, 529.

Hoyle, F. 1948, Mon. Not. R. Astron. Soc. **108**, 372.

Hoyle, F. 1982, Ann. Rev. Astron. Astrophys., **20**, 1.

Hoyle, F. and Narlikar, J.V. 1963, Proc. R. Soc. **A277**, 1.

Hoyle, F. and Narlikar, J.V. 1964a, Proc. R. Soc. **A282**, 184.

Hoyle, F. and Narlikar, J.V. 1964b, Proc. R. Soc. A. **282**, 191.

Hoyle, F. and Narlikar, J.V. 1966, Proc. R. Soc. A. **294**, 138.

Hoyle, F. and Narlikar, J.V. 1969, Ann. Phys. (N.Y.), **54**, 207.

Hoyle, F. and Narlikar, J.V. 1971, Ann. Phys. (N.Y.), **62**, 44.

Hoyle, F. and Narlikar, J.V. 1974, Action at a Distance in Physics and Cosmology (W.H. Freeman, New York)

Hoyle, F. and Narlikar, J.V. 1993, Proc. R. Soc. A.

Hoyle, F., Burbidge, G. and Narlikar, J.V. 1993, Astrophys. J. **410**, 437.

Hoyle, F., Burbidge, G. and Narlikar, J.V. 1994a Mon. Not. R. Astron. Soc. **267**, 1007.

Hoyle, F., Burbidge, G. and Narlikar, J.V. 1994b Proc. R. Soc. **A** (to be published).

Hubble, E.P. 1929, Proc. Nat. Acad. Sci., **15**, 168.

Infeld, L. and Schild, A. 1945, Phys. Rev. **68**, 250.

Narlikar, J.V. 1972, J. Ind. Math. Soc. of India, **36**, 9.

Narlikar, J.V. 1974, J. Phys. **A7**, 1274.

Narlikar, J.V. 1993a, *Introduction to Cosmology* (Cambridge).

Narlikar, J.V. 1993b, in *Philosophy of Science : Perspectives from natural and social sciences*, Eds. J.V. Narlikar, L. Banga and C. Gupta (Indian Institute of Advanced Study, Shimla), p.69.

Padmanabhan, T. 1985, *G.R.G. Journal*, **17**, 215.

Partridge, R.B. 1973, Nature **244**, 263.

Roe, P.E. 1969, *Mon. Not. R. Astron. Soc.* **144**, 219.

Schwarzschild, K. 1903, Göttinger Nachrichten, **128**, 132.

Synge, J.L. 1960, *Relativity : The General Theory* (North Holland)

Tetrode, H. 1922, Z. Phys. **10**, 317.

Wheeler, J.A. and Feynman, R.P. 1945, Rev. Mod. Phys. **17**, 157.

Wheeler, J.A. and Feynman, R.P. 1949, Rev. Mod. Phys. **21**, 425.